P9-DUU-769

WITHDRAWN

Mexico's Petroleum Sector

Performance and Prospect

MEXICO'S PETROLEUM SECTOR

Performance and Prospect

GEORGE BAKER

PennWell Books
PennWell Publishing Company
Tulsa, Oklahoma

For

Weston La Barre
Paul V. Murray
Dan Lundberg
and
John Tate Lanning*
Pamela E. Cull*
Alan E. Walthers*
Betty Baker*
*In memoriam

PennWell Publishing Company
1421 South Sheridan Road/P. O. Box 1260
Tulsa, Oklahoma 74101

Library of Congress cataloging in publication data

Baker, George Towne, 1941—
Mexico's petroleum sector.

Includes index.
1. Petroleum industry and trade—Mexico. 2. Gas, Natural—Mexico. 3. Petroleum chemicals industry—Mexico. I. Title.
HD9574.M6B32 1983 338.2'278'0792 83-4178
ISBN 0-87814-237-1

Printed in the United States of America

1 2 3 4 5 87 86 85 84 83

Contents

Preface

Mexico's oil and gas industry is entering a new phase of its evolution. The six years of unprecedented growth that took place during the presidency of José López Portillo are over. The Mexican economy, heavily in debt, is knocking on new doors to stability and growth. The glamour role given to the oil industry in Mexico during the 1976–1982 period is over, and the future of the industry cannot be projected from any straight line coming out of the past.

Mexico's experience cannot be reduced to an econometric exercise yielding a statistical lesson for other present or prospective oil-exporting nations. Nor should a survey of Mexico's oil industry dwell solely on political, cultural, or psychological idiosyncracies. Little that an American observer says about Mexico can be free of the criticism that it suffers from ethnocentrism at one or another level of analysis.

To try to wrap up Mexico's experience in some cybernetic political or economic model would be to invite a good deal of criticism on this score. What can be said is that Mexicans believe that their system of political and industrial government serves national interests extremely well. The Mexican lock works as if it were designed on American principles, only that no American-made key fits. This failure to fit is why it is so hard, especially at the beginning, for foreigners to succeed in their business objectives in Mexico. The benefit to Mexico from such a system is that national sovereignty and authority remain enigmatically, but indisputably, in Mexican hands.

Mexico does not choose to tell us what lies behind political and statistical appearances, so to take Mexican tables of organization and statistical data at face value is to ask for trouble. For example, Mexico's state oil company, Pemex (Petroleos Mexicanos), each year reports "sales" of petroleum products. However, Pemex means those products that entered the private retail marketplace—Pemex's consumption of petroleum is not included in sales statistics. Calculating the commercial value of Pemex consumption at world prices yields a different picture of the importance—otherwise understated—of the oil industry to the total economic well-being of the nation. However, to create a new model of the Mexican economy is to guarantee that Mexicans will perceive a political agenda behind the analysis.

Mexicans fault foreign suppliers and oil customers for not having an appreciation of Mexico's realities, cultural, political, and economic. However, the implicit invitation to try to understand Mexico's problems and challenges does not extend to an invitation to tell Mexico how to run its oil business.

All of this must be kept in mind when reporting the recent output of the Mexican oil industry. Behind those barrels, thousands of cubic feet, tons, and market share is an analysis, preliminary in places, of the economic costs and incentives associated with this output. However, what is crucial to those dealing with Mexico's petroleum sector is the understanding of how these factors will affect its oil policy and performance in the future. Here, charting Mexico's experience may yield the landmarks of successful business relationships with Mexico in the future as well as with other resource-rich less-developed countries (LDC) in an uncertain energy environment.

Acknowledgments

Mexicans proudly point out that their public officials, unlike ours in the United States, shun trying to make a buck hawking their memoirs after their term of office expires. This attitude helps explain why there are few memoirs of persons in high office in Mexico. With the exception of the memoirs of one director general of Pemex, that of Antonio J. Bermúdez, *Doce años al servicio de la industria petrolera mexicana* [My Twelve Years in the Service of the Mexican Petroleum Industry] (Mexico, 1960), there is virtually nothing in the public domain from Mexican oil men.

Americans may not agree with the Mexican point of view regarding the memoir-writing activities of public officials—perhaps arguing that, on the contrary, it is their duty, implied in taking office, to leave for posterity an accounting of their successes and failures. However, much of the qualitative and quantitative data in the Mexican oil industry is in the hands of persons in Mexico who, following Mexican custom, cannot write for a general or professional audience and for whom it could be awkward to have their names mentioned here.

I am safe, I think, to acknowledge the professional encouragement given to me 10 years ago by Josefina Z. Vázquez, Carlos Bosch García, and Javier Rondero Zubieta, later senator from the State of Morelos. The staff at the Centro de Estudios de la Historia de México, a private research institute sponsored by Condumex, S.A., made available to me the center's unique collection of manuscripts and publications bearing on Mexican history. As for persons inside Mexico's oil industry with whom I have discussed diverse aspects of the industry, I think it is best to follow Mexican custom and omit mention of the names.

In the United States, the situation is only slightly less delicate. Steve Schaffran, German Chacín, and Fernando Arias, among others, helped clarify my thinking about Pemex financial statistics. Abel Beltrán del Río helped me on numerous statistical occasions (and I have used Diemex-Wharton tables as primary data on Mexican macroeconomics). Other persons in the U.S. oil and gas industry, in Houston, Los Angeles and elsewhere, with whom many long-distance hours were spent discussing Mexican oil matters, brought up issues which their respective companies were facing in Mexico—issues that I otherwise might have been unaware of.

World Oil allowed me to draw on work previously published in two essays, one co-authored with Tom Stewart-Gordon, which appeared in the March issues of 1981 and 1982. My approach to analyzing Pemex's income statement was developed in collaboration with *Energy Détente,* and an interpretation of 1975–1980 data appeared in the December 21, 1981 issue.

My associates here in Berkeley contributed a great deal to the manuscript. Jim Culbeaux spent many hours at the computer and plotter in the office. Donald J. Robbins ran regressions and estimates of average and marginal costs. Erik Sivesind, drawing on his background as a resource analyst with experience in the Mexican and Venezuelan oil industries, worked with me very closely in sifting, calculating, and writing Mexican oil sector data. In Marina del Rey, Bob Williams's blue pencil and indispensable whip gave us the animus to complete a manuscript.

Introduction

Mexican oil management is a function of Mexican culture and personality no less than of Mexican history and economics. In taking office in the downward swing of demand in the international oil market, Miguel de la Madrid, who became Mexico's President on December 1, 1982, faces major economic and political problems. Mexican oil during the rule of his predecessor, Jose López Portillo, had transformed itself from backyard significance to world stature.

Mexicans relish and cultivate a business environment of shadowed corners and mysterious reaches—it complicates the game and creates more room for maneuver. The oblique approach is more stimulating for a Mexican, yet more difficult to gauge and react to for a foreigner.

Before embarking on this study of Mexico's present oil industry, some knowledge of the Mexican oil industry's cultural and historical context is indispensable. This type of background information may be skipped by the crude trader or engineering/construction firm in efforts to close a contract. However, if one is to do business in Mexico, or with Mexicans, he must come to terms with the cultural and historical issues on which Mexican business people have been weaned, and which influence their thinking and negotiating positions.

"Mexico," the President of the Republic said, "is not an oil well."

This ironic remark was intended as a reminder to foreign audiences of two related facts of Mexico's past and character: That Mexico has a history that should be taken into account and appreciated by governments and corporations wishing to do business with Mexico; and that Mexico's oil industry, as important as it is from the standpoint of current capital expenditures and revenues, does not occupy such an important place as to make Mexico a standard oil mono-exporter.

However, saying what Mexico is *not* misses saying what Mexico *is*. The Peruvian writer Mario Vargas Llosa said, "[Mexico] is a dictatorship that is periodically renewed, in that the assumed dictator wins the elections cleanly and in that all the elements and ingredients of the system seem to obey its planning with an intensity that is truly uncommon." He characterized the country as having "a margin of democracy that cannot be denied," concluding that despite corruption and other problems, "Mexico is a country that is systematically progressive. It is one of the progressive countries of Latin America."[1]

However, for this novelist, or anyone, to call Mexico a dictatorship, even in a qualified sense, assumes a command of facts beyond the reach of all but an inbred, powerful elite in Mexico whose "200 families" have controlled the country since independence. The country should be put in the class of Black-Box (enigmatic) Republics, a class occupied not only by developing nations but also by some industrial powers such as Japan. It would be safer to err on the side of concluding that nothing much beyond surface detail and columnist gossip is known about the political process in Mexico than to assume that the fundamentals are known and only details are missing.

What can be known by experience and history is something about the Mexican character, some appreciation of his world view, his attitudes toward the outside world in general and toward the U.S. in particular, and his way of doing business. For example, what can one say about a country whose presidential candidate has a graduate degree from Harvard and speaks English fluently, but in public speaks only Spanish and uses an interpreter before English-speaking audiences? What, further, may be imagined about the country if the candidate's official biography mentions only his having studied abroad, with no mention of Harvard or the U.S.? One may infer that the candidate went to Harvard for a sound education yet must downplay this fact fearing that it might tarnish his image as a cultural and intellectual patriot if he were described as "excessively familiar" with Americans. One may merely say of the culture that an ordinary degree of ambivalence toward the U.S. existed—ordinary among developing nations, that is.[2]

In Mexico there is a saying, *Como Mexico, no hay dos,* meaning, "A country like Mexico?—There's only one." But it is not true. Mexico, in most essential political and socioeconomic respects, is like any ordinary, developing country: very proud, very nationalistic, eager to do business with foreigners—but not so eager to please. Marxism-Leninism is as much an intellectual tradition in Mexican universities as it is in universities of other developing countries. Therefore, it is dogma that the rich countries are out to exploit the poor ones. The corollary is that the rich, private corporations within the rich nations are the principal agents of this process of exploitation.

So it follows that any interest on the part of foreign companies in Mexico, be it for product sales or resource acquisition, is suspect. So too, the behavior of Mexicans, particularly those in government who deal with foreign buyers and suppliers, is watched. The fate is sealed of a government official accused of improperly turning over national resources—be they crude oil, timber or tuna—

---

[1]*Excelsior,* 25 January 1982.

[2]See also Appendix K.

to foreigners. This is what happened to Jorge Díaz Serrano, the successful head of the national oil company, who was accused of improperly lowering Mexico's crude oil export price in June of 1981. He was summarily removed from office and within a few months sent to political exile in Moscow—far away from his U.S. business friends in Houston.

In what sense then is Mexico eager to do business with foreigners? The answer is in two words: capital and technology. Mexico is eager to borrow from international banks with a total loan portfolio of about $81 billion outstanding at the end of 1982. The government reasoned—from 1977 to mid-1981, that is—that the price of crude oil would rise faster than the price of borrowing money. Therefore, there was no limit on the desirability of foreign loan capital. Private corporations, meanwhile, found the domestic credit market extremely tight, and dollar loans could be found at nominal interest rates one-half to two-thirds less than those found in peso loans.

Mexico desires foreign investment capital particularly when it brings technology or export market access with it, provided that the foreign company assumes a minority role in the new venture's equity structure. Mexico, like most developing countries, lacks a broad-based product research-and-development tradition. In some areas such as medicine and chemistry, Mexico has a distinguished record but such efforts tend to be government-sponsored.

Private capital in Mexico typically does not believe in spending its own money in substantial research and development budgets. Private capital, moreover, typically does not take export markets seriously. Corporate philosophy is fulfilled when the company is the largest force in a relatively closed, protected market. (The business historian of the López Portillo period may call this the *Alfa Syndrome,* after the stellar rise and eclipse of Grupo Industrial Alfa.) As for investments abroad, Mexicans as individuals buy choice real estate in Texas, California, and Colorado, but it is uncommon to find Mexican companies as shareholders of manufacturing concerns. The result is that Mexican companies are, in principle, interested in joint ventures in Mexico in which foreign partners provide capital, technology, and skills lacking at home.

Mexicans feel best about dealing with foreign buyers and suppliers when there is an established relationship of trust, which Mexicans call *confianza.* Not that such feelings are so different than those of other business people in the U.S. or anywhere else, but in Mexico it is terribly important to search out the unique individual behind the title. That person must know that private conversations will never be betrayed, that the delicacy of the relationship will be understood, and that what is said and done in the name of the title is not to be interpreted as a personal expression.

Having said only this much, it is possible to approach Mexico from the quantitative side of things. It used to be true, and to some extent it still is, that there are never any real statistics for an issue faintly having political overtones.

There are no Mexican numbers, for example, showing the results of sample-population studies of Mexican undocumented workers in the U.S. How many are there likely to be? What is the likely remittance of earnings to the family back home each month? What are the state oil company's annual expenses in dollars? What proportion of Mexico's crude oil reserves consists of light-grade oil? For such questions the Law of Ice (i.e., silence) applies.

Mexicans, in other words, resist giving out data that may be used to compromise future policymaking. Numbers are given out in the most innocuous form possible. The financial section of the annual report of Pemex constitutes an excellent example of Mexican nondisclosure patterns: year's revenues, expenses, gross profits, taxes, and net profits are described in paragraph form with no distinction made between the accounting categories (cost of sales, distribution, etc.) that the same company employs when seeking foreign loans. In the same vein Pemex gives product sales information but with no indication that a given figure represents both the company's official price plus whatever Pemex-billed taxes may apply.

The Mexican soul, of course, is behind all of this. The Mexican, by character, is above all generous; second, courteous and refined; and finally, very much concerned with "face." (Anti-Americanism is one such face found in various pockets in Mexican society.)

The Mexican has the rare trait of being able to laugh at himself, his compatriots, and his culture's unique ways of doing things. He is a master of humor and irony, and anyone doing business in Mexico must be open to this characteristic side of Mexican life. During the period 1970–1976, Mexicans made up thousands, perhaps tens of thousands, of jokes about their president—they even made jokes about their joke-telling. The first few years of the Lopez Portillo period were largely free from presidential jokes. Only at the very end did the sardonic bite of Mexican humor nip at the heels of presidential cuffs.

These are but a few facets of the often enigmatic Mexican character, which derive partly from and reflect Mexican history. A brief look at the development of oil against the backdrop of that history may help reveal some of the rationale behind Mexican oil business motivations to those who are about to enter the maze of Mexico's petroleum sector.

A Capsule History of Mexican Oil

Modern Mexico began with the rule of Porfirio Díaz in Mexico in 1876, marking the beginning of the *Porfiriato,* a period of more than 30 years of rapid, foreign-dominated economic development in Mexico. Centuries before 1876, oil seepages along the coastal plain of eastern Mexico were recorded in Aztec and Mayan histories. The first attempt under Porfirio Díaz to exploit oil came in 1876, at Tuxpan. The enterprising sea captain who sank the first well lost the financing of his Boston partners and committed suicide. This tragic beginning did not discourage subsequent entrepreneurs—the ubiquity of the oil resources was too obvious to ignore.

Although reserve figures are unavailable before 1938, output of oil went from 10,000 bbl in 1901 to 12.5 million bbl in 1911 and peaked at 193 million bbl in 1921, making Mexico the second leading oil producer in the world.

At the time of the nationalization of the oil industry in 1938, proven reserves were less than 1.3 billion bbl of oil equivalent.[3] Reserve figures finally surpassed 2 billion bbl in 1952, and remained at 5–6.3 billion bbl from 1961–1975.

Seepages to gushers—Pre-Columbian to 1900

Since Aztec and Mayan times, the heavily forested, hot, and humid coastal plain of eastern Mexico has been known as a source of oil. From the Panuco River, near present-day Tampico, south to the Yucatán, the presence of oil seepages was reflected in early place names: El Chapopte, El Chapopotal, Chapopotilla, Cerro de la Pez, Ojo de Brea—all meaning tar or pitch. That oil existed along the coast of the Gulf of Mexico was a secret to no one. How much was there and what value it held, however, are questions that have no final answers. A resource is only that when it can be identified. Political, economical, technological, and geological factors determine the quantity and value of a resource such as petroleum.

Some efforts to exploit Mexican oil were made between 1876–1900. The aforementioned Boston sea captain sank some wells near Tuxpan and sold a crudely refined product as illuminating oil. He was but the first in a long succession of foreign entrepreneurs.

The growing importance of oil to industry and transportation, coupled with the favorable political climate in Mexico created by Porfirio Díaz and his *científicos* (scientists), led to the appearance on the Mexican oil scene of foreign interests. Cecil Rhodes formed the London Oil Trust to prospect in the Tuxpan area. After substantial investments failed to yield a profitable return, many of the holdings were farmed out to another British firm, the Mexican Oil Corporation. They also abandoned oil development efforts because of a lack of profit, not for a lack of oil.

Henry Clay Pierce, an American marketing oil in Mexico through his company, Waters-Pierce (an affiliate of Rockefeller's Standard Oil of New Jersey), knew it was there. Pierce had acquired a controlling interest in the Mexican Central Railroad, which ran from El Paso, Texas, to Mexico City to Tampico on the Gulf Coast. He invited Edward L. Doheny, the discoverer of oil in Los Angeles in 1892, to come to Mexico to prospect for oil. Pierce hoped Doheny would find oil near the right of way of the Mexican Central Railroad and promised to buy his production for use as a locomotive fuel.

[3]Reserves always include natural gas in Mexico.

Doheny accepted the offer and headed for the steamy, pest-infested eastern coastal plain and forgot the inclemency and discomfort when he contemplated "this little hill from whose base flowed oil in various directions. We felt that we knew, and we did know, that we were in an oil region which would produce in unlimited quantities that for which the world had the greatest need—oil fuel."

On May 1, 1900, he started drilling at that hill, Cerro de la Pez, and on May 14, tapped into a field that blew the drill out of the ground. Doheny had discovered the Ebano oil field, and the flow from that gusher inaugurated Mexico's oil industry.

From commercial to crucial: 1900–1921

Since the oil resources in those early years cannot be measured in the relatively neat categories now used, such as proven reserves, one can only estimate their extent and importance in the world energy market at the time. By 1905 Doheny's production success and his arrangement with Standard Oil of New Jersey to help supply the New England states had alerted the Mexican government to the possibility of a U.S. monopoly of its oil industry. Other competition was encouraged. Weetman Dickinson Pearson, a British subject later to become Viscount Cowdray, had been exploring since 1901 on the Isthmus of Tehuantepec. Pearson's goal was to establish the first fully integrated oil company in Mexico, which met with limited success. Doheny's dramatic production successes, though, prompted the Mexican government in 1906 to grant Pearson vast oil concessions on government-owned land in five states.

Pearson's exploration efforts completed the job started by Doheny: riveting world attention on the great oil potential of Mexico. On July 4, 1908, Pearson's drillers struck oil in the Dos Bocas well, halfway between Tuxpan and Tampico. The derrick and 4-in. drillpipe were destroyed by the tremendous pressure in the well. The oil ignited, and a 1,000-ft tower of flame shot up and burned out of control for 58 days. This spectacle received worldwide attention from the press and foreign oil interests.

In December 1910 a Pearson rig again hit a gusher in the 4 Potrero del Llano 20 miles west of Tuxpan. Pressures were enormous again, but a fire was avoided. Flow from the well was estimated at 100,000 b/d and finally was capped 60 hard days later. Potrero del Llano was the largest gusher in the world and yielded more than 100 million bbl of oil in 8 years. By the end of 1910, spectacular discoveries and the fact that most wells in Mexico were able to continue steady production under their own hydrostatic and gas pressure had made Mexico exceedingly attractive in the eyes of the world oil industry.

During 1911–1912, Texaco made its first investments in Mexican oil; Gulf made its first foreign investment of any type in entering the Mexican petroleum industry; and Magnolia Oil, under the control of the presidents of Standard Oil of New York and Standard Oil of New Jersey, bought 400 acres of land in Tampico.

In 1901 Mexican crude oil production totaled 10,300 bbl. By 1910 it reached 3.6 million bbl. Then it jumped to 12.6 million bbl in 1911 and to 16.6 million bbl in 1912. Mexico became the third largest producer in the world after the U.S. and Russia.

The Mexican Revolution began in 1910, and its violent period lasted until 1920. Factionalism and endlessly shifting alliances weave many themes through that period. Oil industry concerns played one of the major roles in that protracted conflict, adding to the Mexican oil industry one of its many politicized layers.

The British in 1911, through Pearson's El Aguila company, were in the ascendancy in Mexico's petroleum industry, largely as a result of the backing of Porfirio Diaz. U.S. interests felt discriminated against by Díaz, and several U.S. oil companies—Texaco, Gulf, and Standard of New Jersey—wanted in but could not get concessions from the Díaz regime. Allegations were made and testimony was supplied that U.S. firms provided funds to the leaders of the uprising against Díaz, in hopes of obtaining a more favorable competitive position vis-a-vis the British.

The oil companies were also implicated in alleged activities aimed at overthrowing Madero when it became clear he could not provide Mexican protection for the foreign oil interests. Foreign companies allegedly continued to pay insurgents of nearly every persuasion throughout the revolutionary upheaval. Whether specific allegations are true or not does not alter the fact that both British and American oil interests saw the Revolution as their own battleground for competitive supremacy in the Mexican oil industry. The oil resources of Mexico and investments in Mexican petroleum were great enough that together they comprised the primary concern of both the U.S. and British governments in their reactions to the Revolution.

U.S. oil interests, threatened by warring revolutionary factions, demanded U.S. protection. The threat to those interests was ultimately judged severe enough that U.S. marines landed in Veracruz in 1914 and stayed for 7 months. As Mexicans saw it, oil was at the heart of the major transgressions by foreign, particularly U.S., interests against the sovereignty of Mexico.

Despite all the turmoil of the revolution, petroleum production never faltered, increasing steadily to 193.4 million bbl in 1921 from 16.6 million bbl in 1912. The reason it did not languish under such adverse conditions is because of the meteoric rise of worldwide demand for petroleum. The onset of World War I came on the heels of both the U.S. and British conversion of their navies from coal to oil propulsion. Access to the Middle East (Iran) and Russian oil had been cut by the war. The demands of wartime economies and oil-burning navies could only be satisfied by increased Mexican oil production. During the war, Mexico produced nearly 25% of the world's oil supply. Without it the British navy would have had to revert to the less efficient and slower coal-fired ships or return to the glorious days of sail. Mexico's resources had not only proven their lucrative commercial value and their divisive international power in the first two decades of the century, they also had proven their strategic importance to the Western world.

Apogee to expropriation: 1921–1938

For the foreign companies, Article 27 of the Constitution of 1917, which gave to the Mexican nation ownership of all hydrocarbon resources, continued to be a legal thorn that drew a little blood and plenty of protests. A series of Mexican Supreme Court decisions, denying retroactive application of Article 27 and forbidding the exchange of production and exploration rights for 50-year concessions, combined with some diplomatic compromises to allow the foreign companies to continue their operations. But the Court's ruling did not remove the companies' fear that the status quo could easily change.

During the most tumultuous years of the Revolution, production of Mexico's petroleum increased continuously and rapidly. The resources were there for the taking. Mexican production peaked at 193 million bbl in 1921 and was second in the world from 1918–1927. Ironically, once the violence of the Revolution ended in 1920 and a modicum of political stability returned to Mexico, productivity began a steady decline, reaching 40 million bbl in 1930.

Part of the drop was blamed on depleted oil fields, undoubtedly at least partially true. Continuing political and legal uncertainties concerning ownership of subsoil rights also dampened oil company enthusiasm. Drilling records, however, indicate that the companies increased their efforts to uncover Mexico's petroleum resources after 1920. From 1920–1926 the rate of drilling went up at an average annual rate of 39%. The number of wells drilled from 1924–1927 was more than five times the number drilled in the entire period before 1921. Prior to 1921, only 38% of the wells were abandoned at the end of drilling, while more than 55% were abandoned in the years 1921–1927.[4] These drilling figures seem to refute charges that the companies had written off Mexico as a bad risk.

However, exogenous factors pulled the companies away at the same time that less-successful exploratory efforts and the legal and political uncertainties pushed them away. The rate of drilling dropped after 1926, in part because of the discovery of enticingly large oil deposits in Venezuela and the U.S. At the same time, the wholesale price index for petroleum products in the U.S. dropped 28% between 1926 and 1927 and continued downward until the 1931 index was more than 60% below that of 1926.

The blow-by-blow history of the events that led to the expropriation of foreign oil interests in 1938 has been told many times. Labor presented demands

---

[4] J. Richard Powell, *The Mexican Petroleum Industry 1938–50* (Berkeley: University of California Press, 1956), p. 15, Table B2.

unacceptable to the companies, causing the pro-labor government to set up a hostile commission of inquiry that made an arbitration award favorable to labor. Having developed no *confianza* with President Lázaro Cárdevas and having no sensitivity to the issue of presidential face in Mexico, the companies rejected the terms of the commission's award. In retrospect, as George Philip observes, the only chance for compromise at this point "revolved around the possibility of the companies publicly accepting the terms of the award while a way was found to reimburse them in secret."[5] Leaving the president no room to save face, on March 18, 1938, the companies, as the expression in Mexico goes, *se amanecieron muertas*—"woke up dead." The exploitation of Mexican petroleum resources was now in the hands of the government, where it has remained.

In the chapters that follow three levels of study are undertaken: diachronic analysis, which cuts across time; synchronic analysis, in which Mexico's petroleum industry is seen as an on-going set of commercial, industrial, and political relationships; and, in chapter 12, nomothetic analysis, in which events and institutions in Mexico are treated as "lawful" patterns characteristic of developing countries.

Chapter 1 examines hard-core relationships in Mexico's oil industry. Chapters 2–8, as well as the section of appendices, introduce the principal industrial and financial statistics of the López Portillo period (1976–1982). Chapter 9 explores the constitutional and diplomatic beginnings of U.S.-Mexico petroleum relationships, while chapters 10, 11, and the Afterword look into the future.

Throughout, issues of interest to international managers are emphasized. While the chapters constitute basically a descriptive exercise, the thrust of the Afterword is prescriptive.

---

[5]*Oil and Politics in Latin America* (Cambridge: Cambridge University, 1982), p. 226.

Mexico's Petroleum Sector

Performance and Prospect

1

The structure of the petroleum sector

The question of how Mexico's oil industry—not to be confused with Pemex—is organized can be baffling. Pemex is at the core of the industry, but some of Pemex's activities fall outside the area of the oil industry. Additional actors besides Pemex have important and, in the case of the labor union and the government, crucial roles in the oil industry of Mexico. The oil industry in Mexico is Pemex—its research arm, its suppliers, its distributors, its rulers, and its critics (Fig. 1.1). How these actors are related and where authority and responsibility reside within the structure of the industry are the primary concerns addressed in this chapter.

Several measurements may be used to determine the size of the petroleum industry in Mexico's economy and society. At this point the importance of the petroleum sector in the overall energy picture of Mexico is worth noting. The hydrocarbon sector has gained an overwhelmingly predominant position in Mexico's primary energy supply (Fig. 1.2). In 1970, oil and gas accounted for 73% of total primary energy supplies. In 1980 this proportion had grown to 83.4%. The contribution of different types of hydrocarbon fuels has changed very little from 1970 to 1980. The energy from nonpetroleum-based fuels has dropped its share by about 2% (Fig. 1.3).

Organization

There are two organizational paradigms at work in the energy sectors of developing countries: formal and functional. Describing these two organizational models is partly a matter of intuition, since the topic is extremely sensitive and since top officials don't publish their memoirs.

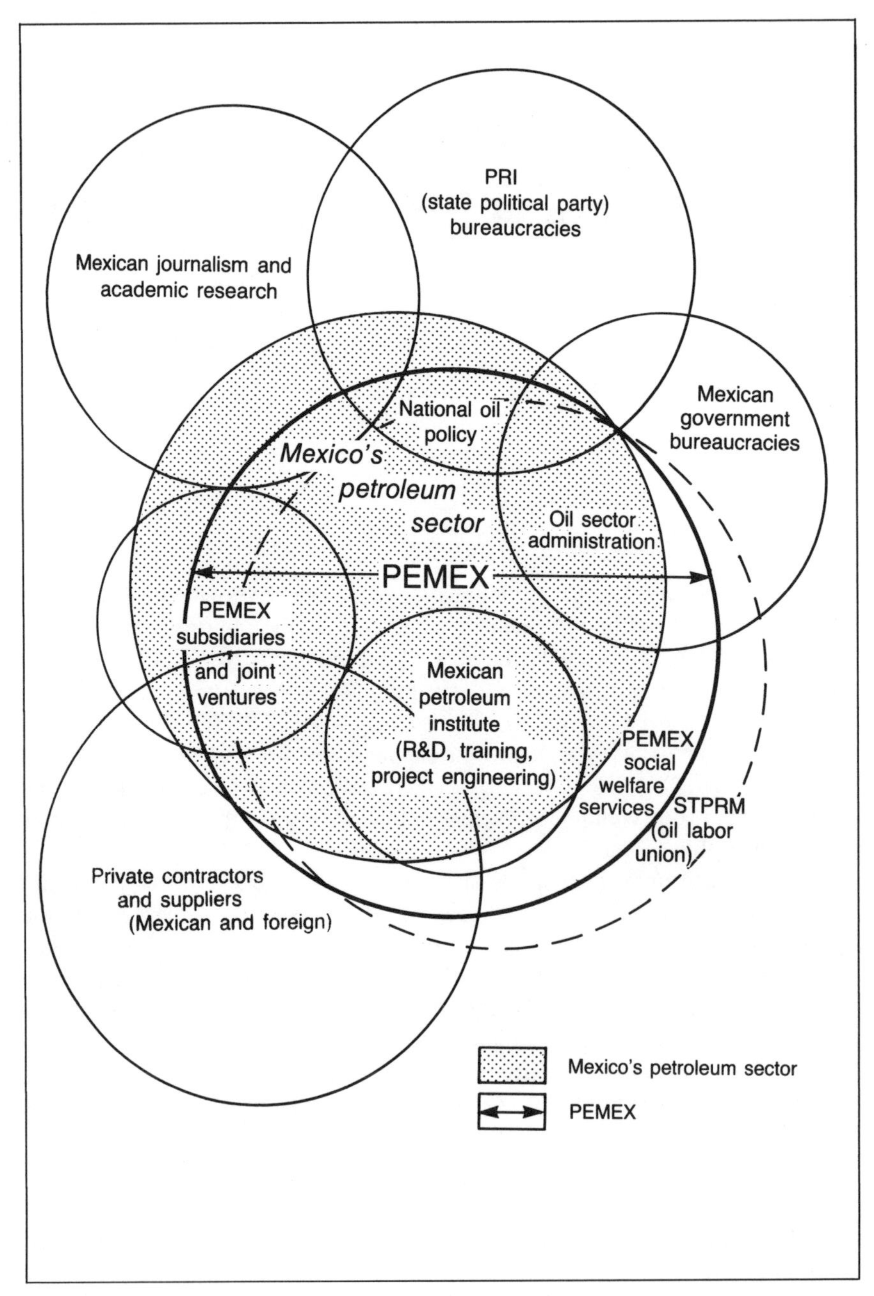

Fig. 1.1 *Who participates in Mexico's oil industry (source: Baker & Associates)*

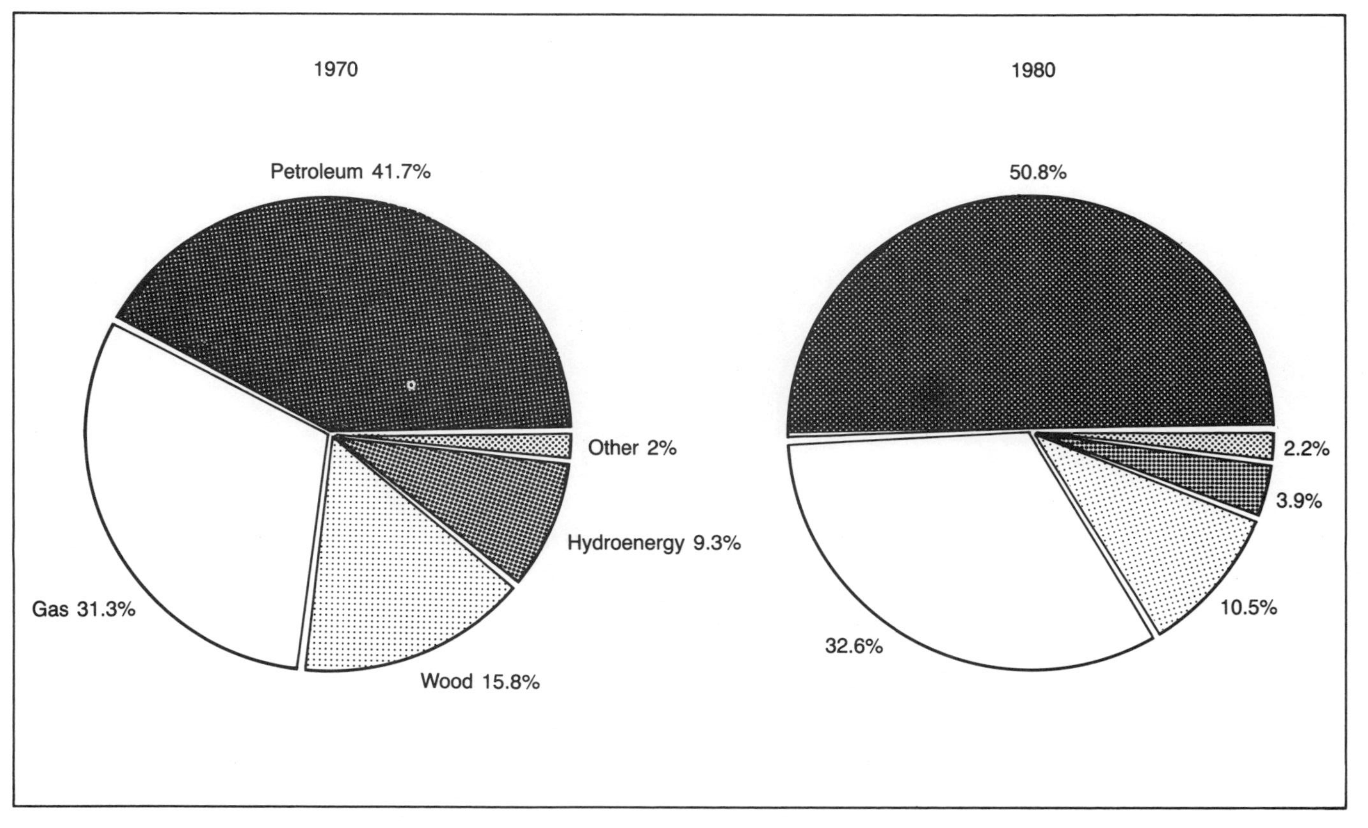

Fig. 1.2 *Petroleum dominates supply (source: Olade)*

Fig. 1.3 *View of secondary energy supply (source: Olade)*

Political and bureaucratic organization of Mexico's energy sector

The organization, which is both formal and informal, exists on two levels: that of the operating company and that of national energy policy. At the operating company level, the two main organizations are Pemex and the Federal Electricity Commission (CFE). The operating companies have responsibilities in the areas of exploration, development, production, storage, and transportation. At the energy policy level, the two principal organizations are the National Energy Commission and the Office of the President.

The members of the Energy Commission under the López Portillo government were the Secretaries of Natural Resources and Industrial Development (changed to the Secretariat of Energy, Mines and Para-State Industry under de la Madrid), Agriculture and Water Resources, Planning and Budgeting, Treasury, and Commerce. The Commission has an executive staff which, organizationally, was subordinate to Patrimonio. In the Office of the President, there are planning and advisory groups monitoring the performance of the energy sector in relation to Mexico's overall domestic and foreign policy goals.

In matters related to the energy sector, the President of Mexico wears three hats. He is chairman of the board of the parent, or holding, company whose subsidiaries are engaged in, or intend to diversify into, all areas of the energy field: hydrocarbons, hydroelectric, nuclear, geothermal, solar, etc. He is chief foreign economic coordinator overseeing the development of bilateral trade, financing, and investment relationships between Mexico and present and prospective buyers, not only of Mexico's crude, but of Mexico's manufactured products, petroleum and nonpetroleum, as well. Finally, as Mexico's chief national security advisor, he proposes guidelines for the strategic protection of Mexico's most valuable international asset, its hydrocarbon reserves.

Formal organization of Mexico's energy sector

A picture of the complex, elusive formal organization of Mexico's energy sector can be diagrammed (Fig. 1.4). There are three levels of organization: 1) the operating companies, 2) the ministry-level managers at the new energy ministry, whose Spanish acronym is Semip, and 3) the presidential economic cabinet and staff.

The operating companies are Pemex (oil and natural gas), CFE (electricity and geothermal), and Uramex (nuclear). Other companies, to exploit solar energy, for example, are not yet capitalized. Semip is principally but not exclusively responsible for the overall management of the energy sector. The operating companies are subordinate to it, and Semip's chief is the formal chairman of Pemex's board of directors. (Also sitting on Pemex's board are the ministers of finance and commerce along with the head of the National Development Bank [Nafinsa] and the CFE.) Semip's Dirección de Energía in Mexico City operates a think tank known as the Technical Secretariat of

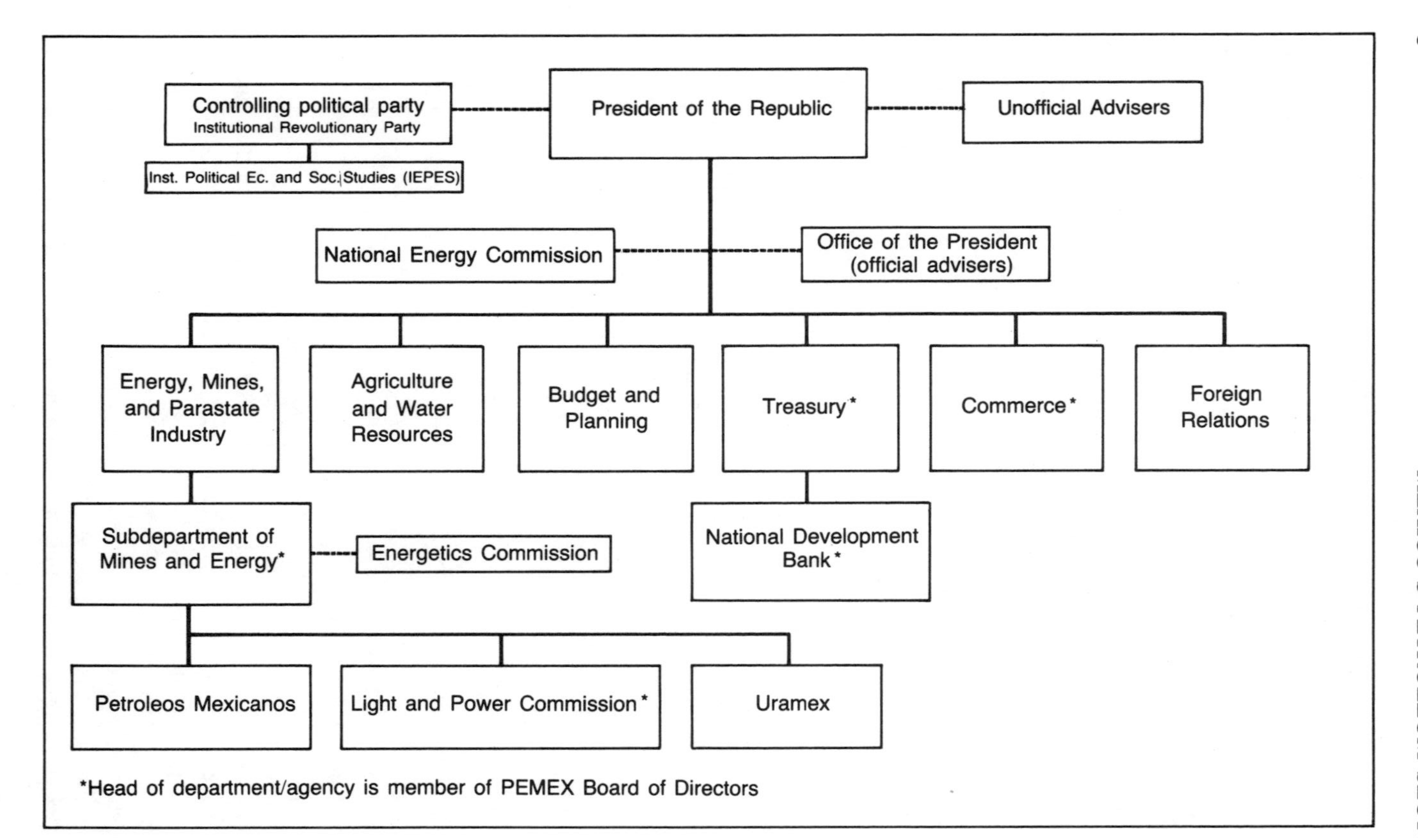

Fig. 1.4 *How Mexico's energy sector is organized (source: Baker & Associates)*

the Comisión Nacional de Energéticos, which has been translated as the National Energetics Commission. This organization produces excellent technical papers on world energy issues and topics, some of them dealing with Mexico.

The Presidential economic cabinet and staff include the heads of those ministries who are most involved in the economic decision-making process in the government. These officials are members of the National Energy Commission, whose staff support is provided by the corresponding organization in Semip.

According to press reports, it was the "economic cabinet" that objected to the crude oil price reduction made in June 1981 by Pemex's director general. The economic cabinet, it was reported, had not been consulted about the price retreat; consequently, the cabinet asked for the resignation of the director general as a symbol of Mexico's opposition to soft-on-OPEC price initiatives.

Unfortunately, Fig. 1.4 gives no clue of the limits of responsibility and authority of any of the organizations appearing in the chart, and for this reason it becomes necessary to construct a functional table of organization for analytic and predictive purposes.

Functional organization of Mexico's energy sector

The functional organization of Mexico's energy sector can be drawn with much cleaner lines than can the formal organization. There are four levels of organization (Fig. 1.5): 1) the operating company with its corresponding energy research institute, 2) affiliates, subsidiaries, and subcontractors, 3) a holding company corresponding to each type of energy source, and 4) a second-level holding company chaired by the president of the Mexican republic.

This approach lets us describe fairly precise limits of authority and responsibility. The second figure differs from the first by translating *comision* in Comision Nacional de Energeticos as "corporation," not "commission," following the precedent inherent in the name of Mexico's light and power company, Comision Federal de Electricidad. *Comision* in this instance indisputably means company or corporation, not "government commission." At the top of Mexico's energy sector, then, is the National Energy Corporation (NEC).

The NEC's principal marketable asset is roughly 80 billion bbl of oil equivalent of crude and natural gas reserves. Although ultimate ownership of these hydrocarbon assets may be said to belong to the citizens of the Mexican republic at large, the NEC has management authority and effective control. Only the NEC, for example, has authority to transfer, at a negotiated price, Mexico's crude oil reserves to the U.S. Strategic Petroleum Reserve (SPR).

The NEC's debt structure cannot be calculated directly. The Mexican government, relying on the marketable value of its hydrocarbon reserves and employing Pemex as the debt-encumbering vehicle, is able to obtain internation-

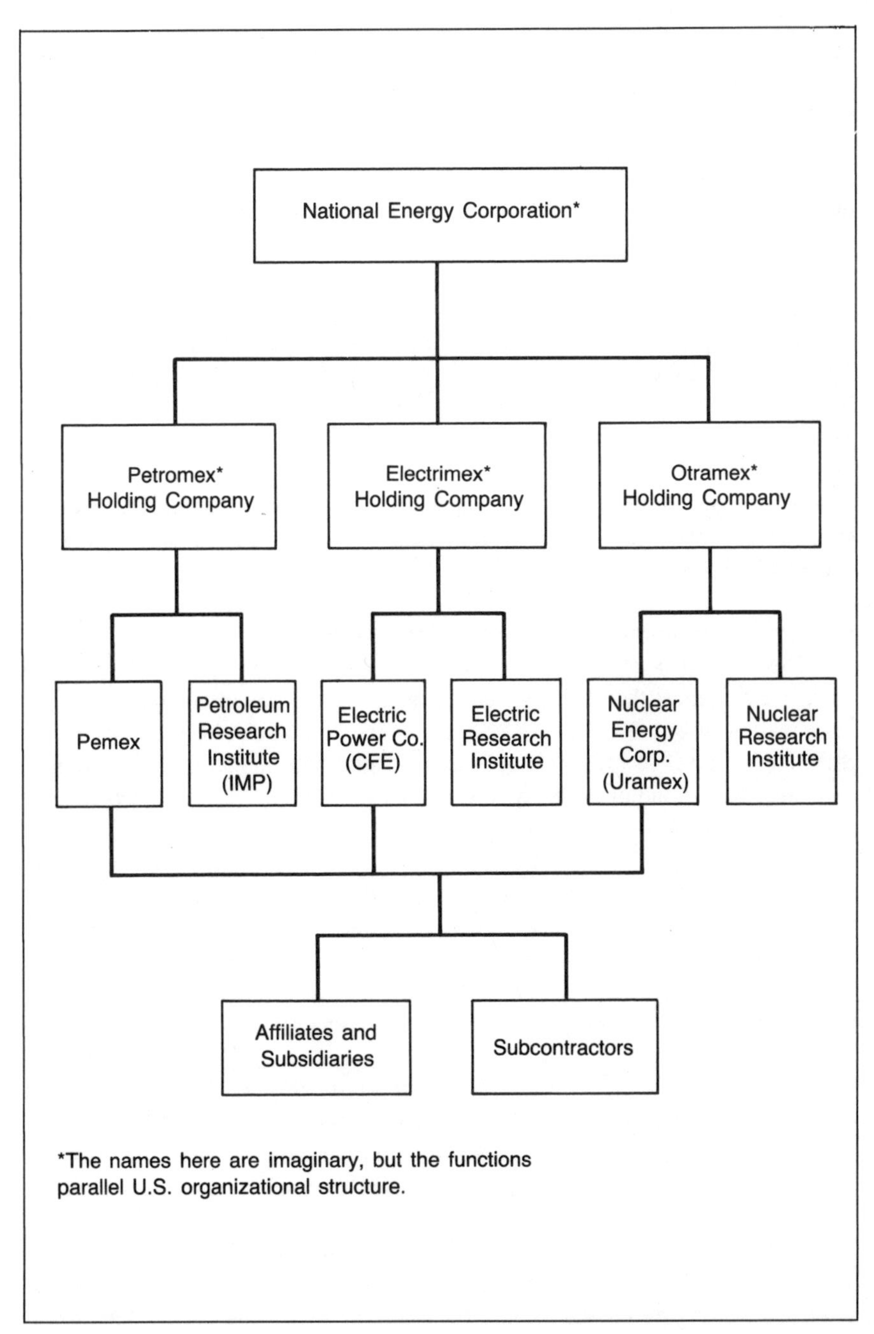

Fig. 1.5 *A look at energy sector's functional ties (source: Baker & Associates)*

al credits in sizeable amounts. In 1981, credit facilities for export purposes were obtained in international capital markets for $4 billion. Such financing arrangements would be unnecessary if Pemex were taxed at the normal 42% for corporations, but Pemex from 1976–82 was taxed at the 97–99% range of gross profits. The difference between 42% and 99% is what Pemex must borrow abroad. At the risk of oversimplification, we can conclude that the NEC is indirectly, through Pemex, borrowing 55–57% of the face amount of any loan obtained by Pemex.

The NEC has a curious personnel system: most of its staff is permanently assigned to the operating or holding companies. The export sales office in Pemex, for example, is staffed by officials who work for the NEC. This organizational relationship can be seen in the handling of the SPR sale. No one in Pemex, and certainly no one at a third or fourth tier of operating company management, would have the authority to conclude any crude oil agreement with the SPR. In this example the sale was made by the NEC, the sales contract signed by Pemex.

Because the NEC's operating income and expenses are not separately reported, for reasons which can be explained, no consolidated income statement for the energy sector can be constructed from financial data currently reported.

The NEC functions in several areas of critical importance to Mexico's energy sector: export sales of crude oil, domestic sales of all energy products, public and industrial relations, executive personnel, investment policy, and foreign aid. These functions can be illustrated by reference to the NEC's performance in the oil subsector.

The crude oil export market diversification policy is the work of the NEC, not Pemex. NEC officials, most of whom receive salaries from other organizations, arrange for Mexico's government-to-government oil deals. These agreements typically include some reciprocal obligation which the buyer of Mexican crude undertakes in exchange for the oil supply contract.

Domestic energy price policy for oil, natural gas, and electricity is decided by the NEC, not by the operating companies. Historically, these prices have been well below those of the international market. For this reason, profit and loss responsibility should be thought of as belonging to the NEC, not to the operating companies themselves. It would be a simple decision by business managers in the operating companies to raise domestic prices so that income would exceed outlays by respectable margins, but operating company managers do not have such authority.

Semip's president serves as vice-president, public and industrial relations, for the NEC. It is his job to defend both popular and unpopular decisions before domestic and international audiences. It was his task during the walkout by crude oil customers in June and July of 1981 to threaten them with being purged forever from Mexico's list of preferred customers.

The finance minister is chief financial officer of the NEC, and his recommendations help decide the level of the public sector indebtedness to be incurred by the implied collateral of crude oil reserve assets.

The president of the country is probably the person who doubles as vice president of executive personnel of the NEC, although it is highly probable that he is assisted by leaders from his party. This position requires him to appoint and remove officials from both the operating companies and the ministries. Policy and personnel decisions of the NEC are transmitted through the holding companies to the operating companies.

As a contract supplier to the federal government, subordinate to the NEC, Pemex has authority limited to industrial operations. Pemex has no independent authority in the areas of marketing, sales, and corporate strategy. It was for this reason in June of 1981 that the president of Pemex was removed from his position for having, on his own initiative, telexed foreign buyers that the average price of Mexican crude for the third quarter would be reduced $4/bbl. While the soundness of this pricing decision was later upheld by the marketplace, the unsolicited intervention of Pemex in national oil export prices was sufficient cause for disciplinary action.[1]

It is as if Pemex is paid on a sales commission basis from the federal government but, for purposes of financial reporting, Pemex's sales revenues are ascribed to itself and payments to the government appear as taxes. And, for most purposes, this manner of describing Pemex's operations and financial condition is satisfactory. It becomes unsatisfactory where we want to evaluate the management performance—not only of Pemex as a diversified oil industry services company, but of the energy sector management organization above Pemex which, since the fall of 1980, has increasingly been involved in the day-to-day operations of Pemex.

Some indication of the NEC's management performance, however, is possible. During the period 1976–1981, NEC-orchestrated developments in Mexico's oil industry included the following:

- Increased export sales revenue from 1976 to 1981 at an annual rate of 115% in current dollars, thanks to a largely successful policy of OPEC-plus pricing.
- Successful diversification of crude exports, bringing the U.S. share of Mexican crude exports down from 90% to 50% in 1981.
- The negotiation of a regional energy program, the Pact of San Jose, in 1980 with Venezuela and Central American and Caribbean countries.

---

[1]A precedent for Díaz Serrano's action existed, but in reverse: Bermudez asserts that in 1958 on the last day of his tenure as head of Pemex, he unilaterally ordered domestic oil product prices to be raised (from Philip's *Oil and Politics in Latin America,* Cambridge, 1982, p. 339).

- A substantial reduction, in real terms, in the price of domestic petroleum products, a trend reversed at the end of 1981 and throughout 1982, with major increases in domestic gasoline and diesel prices.
- The initiation of natural gas exports to the U.S. in 1980 after a 2-year delay owing to the collapse of sales negotiations in 1977.
- The development of a long-range energy program designed to rationalize the production, consumption, and export of nonrenewable energy resources.
- The initiation of talks with Venezuela and Brazil about the formation of a jointly owned international oil company, Petrolatin.

These developments took place in the orbit of national economic and security policy, and responsibility for them lies with Mexico's president and his economic cabinet. In some of these developments Pemex personnel served as agents of and consultants to Mexico's president—in export sales matters, for example. But in such instances the functions performed were on behalf of the president and were not discretionary matters in which, as a Pemex official, an individual was at liberty to take independent action. No Pemex official, for example, had either the responsibility or authority to modify Mexico's crude oil prices.

With regard to organizational theory, it is understandable why, in Mexico's case, the NEC's central role in the government's management of the energy sector is underplayed: Pemex is Mexico's symbol of autonomy not only in matters related to energy but in matters related to technology, industrial development and political self-definition. For this reason, the NEC must be kept in the background and the myth of Pemex's preeminence must be preserved.

But from an outsider's point of view, a satisfactory explanation of the management behavior of the energy sector can be obtained only through business decision models such as the one outlined here.

Price structure of the energy sector

While the international price of energy soared in the 1970s, in Mexico things remained quiet. True, the old, old days—when the peso was 12.5 to the dollar and the price of diesel fuel was 30 centavos/liter (9¢/gal)—ended in December 1973 when both diesel and gasoline prices were raised. However, while the export price of Isthmus crude went to over $35/bbl in 1981, the same barrel of oil in the domestic market stayed in the $7–9/bbl range (Fig. 1.6). While in 1973 domestic oil sold at home for over twice its export value, by 1981 this relationship had fallen to less than 30% (Fig. 1.7).

Data for 1981 illustrate structural patterns in the López Portillo period. In that year Mexico produced 3.65 million barrels of crude oil equivalent (BOE)

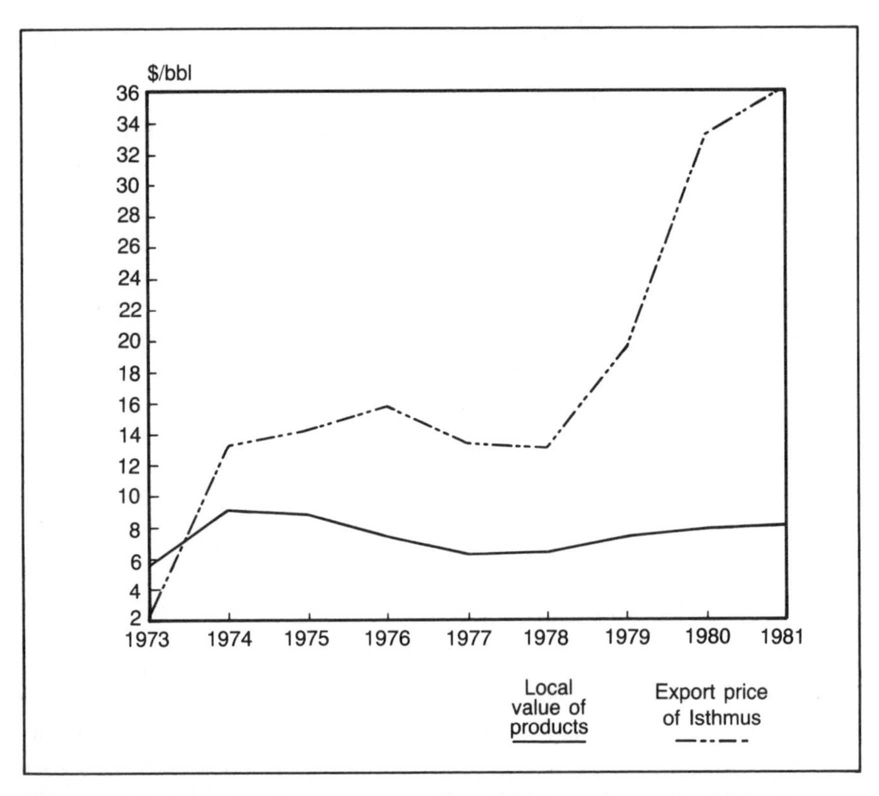

Fig. 1.6 *Refined product revenues flat (local market value/bbl), (source:* Energéticos, *January 1982)*

per day of energy. Of this amount 1.29 million BOE/day were exported as crude oil (1.098 million b/d) and natural gas (302.5 Mcfd). This left 2.36 million BOE/day as gross domestic energy supply.

But, of this latter amount, only 74% was sold on the commercial market. Of the remaining 26%, 10% represented losses (0.247 million BOE/day), and 16% constituted noncommercial government consumption (0.367 million BOE/day).

Total domestic energy sales of petroleum products, gas, and electricity (636 million BOE) scaled by total revenues from domestic sales ($5.88 billion) is the average price of commercial energy. This price was $9.24/BOE in 1981. This price falls to $6.83/BOE, however, if revenues are scaled by available domestic supply (861 million BOE). The difference between these two prices represents noncommercial consumption, the main upstream subsidy given by the Mexican government to the energy sector. The downstream subsidy would be that given to the commercial consumer, represented as the weighted difference between domestic and international prices. In 1981 the average upstream

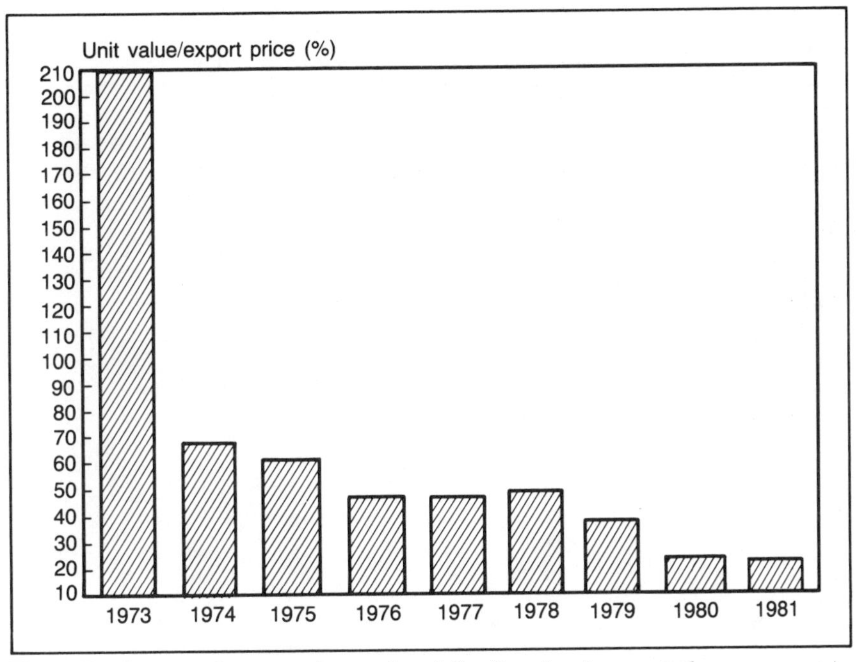

Fig. 1.7 *Income from product sales falls (local sales vs Isthmus exports), (source:* Energéticos, *January 1982)*

Table 1.1
Average price of Mexico's commercial energy in 1981

	Revenue (Million $)	*Domestic sales volume (Million BOE)*	*Net national production (Million BOE)*	*Sales/net national production (Percent)*
Petroleum products	2,972.81	403.36	442.84	91.08
Natural gas	272.66	95.3	256.27	37.19
Electricity	2,637.6	137.73	162.72	84.64
Total	5,883.07	636.39	861.83	73.84

A. Average national price of commercial energy:	
$/BOE	9.24
$/1,000 cu ft	1.71
¢/kwh	2.22
B. Revenue per unit of net national production (gross less exports):	
$/BOE	6.83
$/1,000 cu ft	1.28
¢/kw h	1.64
C. Average upstream subsidy (A – B):	
$/BOE	2.41

Source: *Energéticos* (April 1982) for 1981 sales data.

Table 1.2
What energy products sold for in Mexico in 1981

Energy product	*Revenues (Million $)*	*Volume (Million bbl)*	*Caloric value (Million kcal/bbl)*	*Volume (Million BOE*)*
Total production less exports		407.92	1.3921	442.84
Sales:				
Gasoline	1421.25	130.597	1.2957	131.96
Kerosene	412.68	24.418	1.4057	26.77
Diesel	536.01	85.008	1.4696	97.42
Residual fuels	307.49	91.104	1.593	113.18
LPG	295.38	41.5	1.0515	34.03
Total	2972.81	372.627		403.36
		Average prices		
	$/bbl**		7.98	
	$/BOE		7.37	
		Revenue from net production†		
	$/bbl		7.29	
	$/BOE		6.71	

*Taking 1,282,314 kcal/bbl of oil equivalent.
***Energeticos* (January 1982) calculates that in 1981 Pemex received $8.05/bbl of revenue per barrel of domestic product sales; this would include nonenergy products such as asphalt.
†Gross production less exports.
Source: Pemex, *Memoria de labores, 1981; Energeticos,* April 1982; caloric values from Secretaria de Patrimonio y Fomento Industrial, *Programa de Energia: Balances de energia y estadisticas complementarias, 1981,* p. 136.

subsidy given to the energy sector was $2.41/BOE (Table 1.1). The average price of petroleum products, meanwhile was $7.37/BOE (Table 1.2).

The situation with natural gas was a little different: slightly less than 42% of gross production was sold, 35% at home and 7% in the U.S. Total revenues from natural gas sales in 1981 were $273 million, which, expressed in crude oil equivalent prices, was $2.86/BOE (Table 1.3). The commercial status of electricity in Mexico in 1981 fared much better than did natural gas: the weighted price in crude equivalent terms was $19.15/BOE (Table 1.4). Petroleum products and natural gas sold on the average below the average price of commercial energy, while electricity sold substantially above it (Fig. 1.8).

These patterns would change substantially if each energy source in the domestic market were required to be financially self-sufficient. For 1983 this would require an additional $6.6 billion in domestic sales revenue. One of the

Table 1.3
What Mexico got for its natural gas at home in 1981

	Volume		*Caloric value*	*Barrels of oil equivalent*
	(MMcf)	*(Million cu m)*	*(Trillion kcal)*	*(Million BOE)*
Gross production	1,482,196	41,971	355.08	276.90
Flaring	242,784	6,875	58.16	45.36
Shrinkage,	150,594	4,264	36.08	28.13
Condensation, line rupture	12,293	348	2.94	2.30
Other loss	8,510	241	2.04	1.59
Net production	1,068,015	30,243	255.86	199.53
Exports	110,430	3,127	26.45	20.63
Statistical errors	4,369	124	1.05	0.82
Net available domestic supply	953,216	26,992	228.35	178.08
Pemex consumption	452,107	12,802	108.31	84.46
Sales*	510,109	14,445	122.20	95.30
Sales revenue (Million $)	272.66			
Revenue/unit sale				
$/Mcf	0.53			
$/BOE	2.86			
Revenue/gross production				
$/Mcf	0.18			
$/BOE	0.98			
Revenue/available domestic supply				
$/Mcf	0.29			
$/BOE	1.53			

*Pemex's *Anuario estadistico 1981* gives 14,629 million cu m for sales in 1981.
Source: Pemex, *Memoria de labores, 1981;* conversion factors as in Table 1.2

great debates in Mexican policy circles, however, is the argument over price *rationalization*. To rationalize means to settle on what is to be quantified and how, but little agreement exists in Mexico in the areas of domestic energy prices and subsidies. Much of what is quantified is done so in a highly nationalistic way. Some of these quantifications can be decoded. An initial step is the proper translation of Mexican code words.

Table 1.4
Calculating oil equivalent electricity rates in Mexico
1981

Sector	*Commercial sales* *(Terawatt-hr)*	*(Million BOE*)*	*Rate* *(¢/kwh)*	*Revenues* *(Million $)*
Gross electricity production	67.879	162.72		
Consumption by electric sector	1.995	4.78		
Transmission loss	8.766	21.01		
Sales				
Industrial	31.934	76.55	3.77	1,203.91
Agricultural	3.842	9.21	3.77	144.84
Subtotal	35.776	85.76		1,348.76
Residential	11.211	26.88	6.06	679.39
Commercial	6.125	14.68	6.06	371.18
Public	3.932	9.43	6.06	238.28
Subtotal	21.268	50.98		1,288.84
Other	.41	0.98	NA	—
Total	57.454	137.73		2,637.60
	Average prices			
	¢/kwh	4.59		
	$/BOE	19.15		
	Revenue/unit of gross production			
	¢/kwh	3.89		
	$/BOE	16.21		

*Taking as conversion factors 859,845 kcal/kwh and 1,282,314 kcal/bbl of oil equivalent (BOE)
Source: *Energeticos,* April 1982; conversion factors from Secretaria de Patrimonio y Fomento Industrial, *Programa de energia: Balances de energia y estadisticas complementarias, 1981,* p. 136.

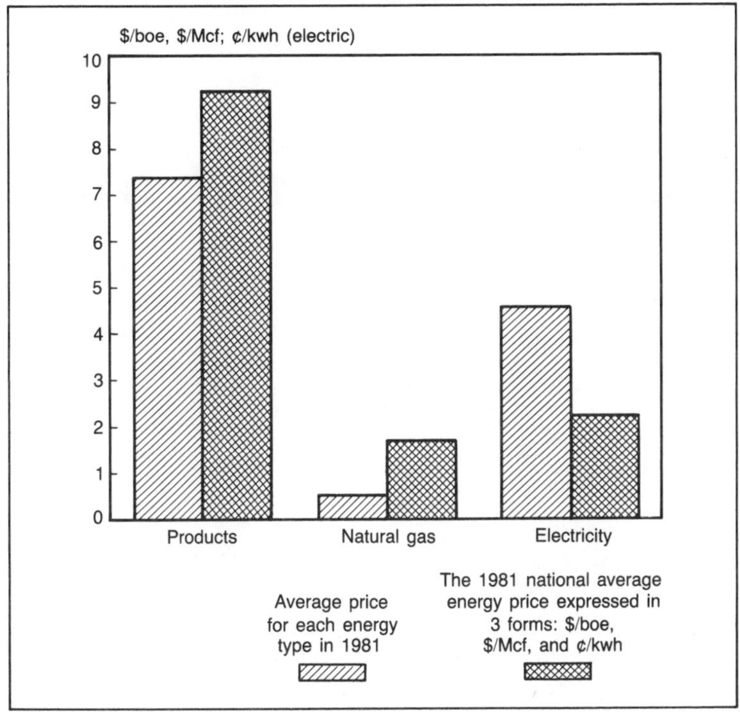

Fig. 1.8 *Electricity overshoots national average energy price (source: Tables 1.1–1.4)*

Table 1.5
Average price of Mexico's commercial energy in 1981

	Revenue (Million $)	*Domestic sales volume (Million BOE)*	*Net national production (Million BOE)*	*Sales/net national production (Percent)*
Energy products	2,972.81	403.36	442.84	91.08
Natural gas	272.66	95.3	276.9	34.42
Electricity	2,637.6	137.73	162.72	84.64
Total	5,883.07	636.39	882.46	72.12

Average national price of commercial energy:

$/BOE	9.24
$/Mcf	1.71
¢/kwh	2.22

Revenue per unit of net national production:

$/BOE	6.67
$/Mcf	1.23
¢/kwh	1.60

Source: Tables 1.2–1.4.

2

The size and shape of Mexico's oil sector

During the period of the López Portillo government, several conventions of speech held sway. One was that Mexico was not an oil exporter, as the term is usually understood in Mexico; rather, it was a country with oil in excess of national demands. Foreign journalists were often taken to task for failing to understand the significance of this distinction.

Quantifying energy subsidies

A second convention was that domestic prices of petroleum products, which were low compared with international averages, did not constitute subsidies. But Mexican prices reflected the cost of production and distribution (read: selling costs) and showed the comparative advantage that Mexico had over less-fortunate economies because of its wealth of oil. Thus—so went the argument—Mexico did not "subsidize" the products of its government-owned company. Not until mid-1982 did one hear Mexican government officials complaining in public about the so-called subsidized prices of petroleum products in the national market. (This convention of speech was to be abandoned by the de la Madrid team, which in the National Development Plan for 1983–1988 spoke not only openly but proudly of the energy sector's contribution to economic growth by means of its sales to the internal market "at subsidized prices.")

The third convention had to do with the size of the participation of Mexico's oil industry in the economy. Mexico wanted it plainly known that the country was not an oil economy like that of the Middle Eastern oil producers. The highest politically acceptable number was 6%: Mexico's oil industry constituted, at most, only 6% of the national economy. The numbers and argument that were used to make this case were striking (Table 2.1).

Table 2.1
How oil officially figures in Mexico's economy

	(Billion current pesos)					
	1976	*1977*	*1978*	*1979*	*1980*	*1981*
Gross Domestic Production (GDP)	1,228.0	1,674.7	2,122.8	2,767.0	3,824.3	5,190.2
Product of petroleum industry	37.4	64.1	83.1	132.4	208.9	289.6
Petroleum total production (%)	3.0	3.8	3.9	4.8	5.5	5.6

Source: Diemex

These calculations erred, however, in taking Pemex numbers for fair market values. Pemex numbers are distorted by the effect of any and all subsidies touching on the oil sector of the economy. One subsidy under President Luis Echeverría Alvarez and later López Portillo was the exchange rate. Pemex's accounts are credited in pesos for dollar-denominated exports. Where an exchange rate subsidy was in place, Pemex received too few pesos for the dollar revenues generated. If Pemex reported a barrel of oil exported at $32 at a period in which the exchange rate was 25:1, then the company's accounts were credited with 800 pesos (25 × 32). However, if the truer exchange rate, based on parity of purchasing power (netted of inflation), were 45:1, then Pemex's accounts should have been credited with 1,440 pesos (45 × 32). Pemex's revenue accounts, in this example, would have been understated by 44%.

Another sort of subsidy affected domestic sales. Inside Mexico Pemex traditionally sells petroleum products well below international prices—even in the northwestern states, importing products from U.S. refineries (such as motor fuel from Union Oil Company) and selling them *at a loss*. However, to adjust Pemex's figures for domestic sales requires that two issues be settled. One issue is the amount of the subsidy in each year to motorists on both sides of the U.S.-Mexican border buying Mexican gasoline and to industrial consumers who buy diesel, residual fuel, natural gas, and petrochemicals at below-world prices.

The amount of each product's subsidy can be calculated by comparing Mexican prices to international prices.[1] Alternatively, the aggregate effect of price subsidies can be estimated by comparing Pemex's domestic product revenues to the revenue that would have been generated had domestic petroleum products been valued at crude export prices. By adding the value of this subsidy to the official figure for the contribution of the petroleum sector to the economy (given in Table 2.1), a new set of figures emerges. The revised figures for the oil

[1]See also Appendix D.

sector, when shown as a percentage of GDP, are considerably higher. Officially, Mexico's oil sector was 5.5% of GDP in 1980; however, if the value of the subsidy is included, the sector's contribution was nearly double that level (Table 2.2).

The other side of the ratio of Pemex sales/national income can be considered by asking what the status of energy subsidies is in Mexico's national income accounting. In López Portillo's Mexico, energy subsidies served as a mechanism for income redistribution and as a policy instrument to channel private investment. Suppose the rack price of a gallon of gasoline was $1.00 in the U.S., but 50¢ in Mexico. A U.S. refinery would have received $1.00 in revenue for each gallon sold, while Pemex received half that amount. Concluding that the fair market value of the product was $1.00, one could argue that Pemex management, acting rationally and prudently, in effect sold the product for $1.00, but at the government's request simultaneously transferred to the consumer an economic subsidy worth 50¢. The implication is that Pemex's sales accounts should be revised upward to show the true market value of products sold in the domestic market and that government subsidies should not be allowed to remain hidden in Pemex sales figures.

Table 2.2
Estimating Mexico's domestic price discounts

	(Million bbl or billion current pesos)					
	1976	*1977*	*1978*	*1979*	*1980*	*1981*
Total crude production*	292.3	358.1	442.6	536.9	706.7	843.9
Crude export volume	34.4	73.7	113.2	194.5	302.1	400.8
Crude volume charged to domestic consumption	257.9	284.4	309.4	342.4	404.6	443.1
Crude export revenues	7.0	23.4	41.8	91.7	221.9	340.8
Export market value of crude charged to domestic consumption	52.5	90.4	97.1	161.4	297.2	376.8
Actual domestic product revenues**	38.4	52.8	59.1	74.6	95.4	112.1
Value of implied subsidy	14.1	37.6	38.0	86.8	201.8	264.7
Revised product of petroleum industry	51.5	101.7	121.1	219.2	410.7	554.3
Subsidy as % of export market value	26.9	41.6	39.1	53.8	67.9	70.2
Petroleum industry product (revised) as % of GDP	4.2	6.1	5.7	7.9	10.7	10.7

*Includes condensates.
**Includes petrochemicals and natural gas.
Source: Table 2.1; Banamex, *Mexico Statistical data, 1970–80;* Amcham, *Energy Mexico, 1981,* and *Memoria de labores 1981.*

However, if the energy subsidy is taken out of Pemex's accounts, there are several places in Mexico's national income accounts that it could be distributed. Is this subsidy more like a retirement pension, in the sense of being a payment for prior services, or is it more like a payment to an employee or consultant for services currently being rendered? In the former case the 50¢/gal economic coupon that the government hands out to the consumer at the point of purchase—resulting in his having to pay only the remaining 50¢ for the gallon of gasoline—is to be treated as a simple transfer of funds or redistribution of income, not as a government expenditure for goods and services.

On the other hand, if the energy subsidy is a payment for current productive services, it should be treated as a cost. Under López Portillo special energy subsidies were offered to industrial consumers who would establish new facilities in out-of-the-way places marked for regional priority development. Such a policy permits the assumption that the government believed its subsidy was payment for current services rendered, in this case, "regional development services."

Clearly, not all of the value of the subsidy was conceived of as payment of services, and none of it was entered into national income accounts. Because the government reasoned that it could set its energy prices as it chose, the difference between those prices and international prices was not entered into national income calculations as payment for services and, hence, as government consumption. Calculations of government consumption and national income, therefore, should also be adjusted upward to reflect the value of the energy subsidy.

This argument suggests that to calculate the contribution of Pemex to national income in a given year in percentage terms, the denominator (national income) must be increased by the value of all petroleum energy subsidies that paid for services rendered to the government. Further, the numerator (Pemex sales) must be increased. For exports Pemex's accounts should be adjusted for the effect of an overvalued Mexican currency depressing peso income from dollar-denominated sales. For domestic sales Pemex's accounts should be adjusted for the effect of the government's mandatory rebate program that depressed recorded income figures (Table 2.2). Having made these adjustments, the contribution of Pemex's operations to national income can be recalculated. A full measure of the contribution of Mexico's oil industry is not yet possible, owing, in part, to a problem of definition.

Definition of oil sector

Before turning to these calculations, some effort at defining Mexico's "oil industry" is called for. In Mexican government publications, "la industria petrolera" is synonymous with the functions and activities of the state oil company, Pemex. An extremely useful compilation of data, *La industria petrolera en Mexico* (1980), for example, speaks of the participation of the petroleum industry in the total imports of Mexico, noting in a footnote that the term "petroleum industry" refers exclusively to Pemex. Such a definition understates the importance of Mexico's—for lack of a better term—greater petroleum industry. The "greater" industry includes not only Pemex but all those other firms, government and privately owned alike, that supply goods and services to Pemex and to each other. It also includes all those government agencies and officials who, on a full-time or part-time basis, perform research, management, or public relations services for Pemex and Mexico's image as an oil exporter, as the term is politically and ideologically understood.

The greater industry even overlaps the formal Pemex structure. Retail gasoline dealerships, all of which sport Pemex signs by their driveways, are in fact almost all privately owned. In 1982, the government recognized the Confederation of Gasoline Distributors of Petróleos Mexicanos as the dealers' exclusive representative in price and commission negotiations. Pemex distributorships used to be part of the patronage system in Mexico and were given to deserving government and union officials.

A definitional problem exists here because the "oil and gas industry," as the term is understood in the U.S., certainly does not include the value of the services performed on Capitol Hill by the Joint Committee on Energy or in the miniature energy departments that have propagated themselves in the capitals of a number of states. On the contrary, such service, if accounted for at all, would be thought of more as an expense.

The "industry," understood as private-sector activity, has institutionalized this expense through heavy contributions that fund the counterservices that the American Petroleum Institute (API) in Washington provides. The API's mission is, in part, to resist government actions deemed harmful to the petroleum industry through public relations, education, and lobbying activities. An arguable point might be whether the API is part of the U.S. oil and gas industry. If the definition given above is used—all private sector activities, whether or not established on a profit basis—then API and other not-for-profit institutions are included in a definition of the oil industry. American business and legal traditions make a sharp distinction between government services in the petroleum industry and in private industry.

All of this changes when you go to Mexico. The oil industry in the eyes and statistical tables of the government is Pemex taken by itself. Government statisticians, as in the U.S., do not include themselves as part of the industry they are reporting on. Pemex, however, is government owned. There is no distinction at the level of equity between Pemex and the rest of the government.

The rest of the government, meanwhile, contains numerous departments, offices, and officials who are assigned, at least part time, to oil sector activities. An example is the Mexican Petroleum Institute (IMP), which is established as a research, training, and engineering consulting organization separate from Pemex but whose principal client is Pemex itself. The IMP is often given the prime contract by Pemex to do project engineering work; the IMP, in turn, is free to hire local as well as foreign subcontractors. Pemex may in this way indirectly import a good deal of intermediate goods and services that would not appear on its records.

One solution to the definitional problem of Mexico's petroleum industry is to use the term "petroleum sector," one that is commonly used in Mexico, to include Pemex, the IMP, prorated parts of the government agencies such as the Energy Ministry (and its Technical Secretariat) devoted to the oil and gas industry, and those private firms—Mexican and foreign—who manufacture secondary petrochemicals or who provide capital goods or goods and services to Pemex and the IMP. The term "petroleum sector" should also include the output of the monopolistic labor union Sindicato de Trabajadores Petroleros de la Republica Mexicana (STRPM), which controls Pemex workers and, it is often alleged, traffics in Pemex jobs.

Mexico's Resource Base

Historically, the primary source of Mexican oil has been beneath the Poza Rica (Wealthy Well) district in the Central Zone of the country's exploration areas. Since 1970, reserves in the Central Zone have been declining both absolutely and as a percent of total reserves. In 1970, the Central Zone accounted for 41% of total hydrocarbon reserves. By 1981, its contribution had dropped to only 3% of proved reserves. In contrast, the Southern Zone in 1970 held 17% of total reserves but by 1981 constituted 68% of reserves.

Clearly, the epicenter of Mexico's reappearance in the world of oil lies beneath the surface of the Southern Zone. Within three years of Pemex's major Reforma discovery in 1972, total proved reserves went from 5.7 billion bbl to 11.2 billion bbl at the end of 1976. By the end of 1981, proved reserves totaled 72 billion bbl after an average annual growth rate of 45.1% from 1976 to 1981. The Southern Zone led this rise with an average annual growth rate of 51.3% over the same period. The Southern Zone has the dominant role in Mexico's overall reserve picture (Table 2.3).[2]

Southeast Mexico

Because of the preponderant importance of reserves in the Southern Zone (the Reforma area and Campeche), that area has become the focus of studies attempting to estimate Mexico's hydrocarbon resources.

The large potential of southeastern Mexico is associated with a giant barrier reef formed along the ancient Yucatan platform during the Cretaceous-Jurassic

---

[2]*Oil & Gas Journal* (30 August 1982), p. 95.

Table 2.3
How Mexico's official reserves soared

	1976	*1977*	*1978*	*1979*	*1980*	*1981*
			(Million bbl oil equivalent)			
Total hydrocarbons	11,161	16,002	40,194	45,804	60,126	72,069
Northern Zone	2,353	3,073	3,505	3,182	2,763	2,977
Central Zone	2,614	2,616	2,561	2,509	2,526	2,362
Chicontepec	—	—	17,640	17,608	17,604	17,597
Southern Zone	6,195	10,313	16,489	22,504	37,234	49,133
			(Million bbl)			
Oil	7,279	10,428	28,407	33,560	47,224	56,998
Northern Zone	768	821	801	690	731	727
Central Zone	1,890	1,887	1,813	1,758	1,723	1,606
Chicontepec	—	—	12,285	12,261	12,257	12,252
Southern Zone	4,621	7,720	13,508	18,852	32,512	42,414
			(Million cu ft)			
Natural gas	19,410	27,868	58,935	61,217	64,511	75,352
Northern Zone	7,924	11,260	13,519	12,462	10,159	11,248
Central Zone	3,619	3,648	3,740	3,756	4,013	3,782
Chicontepec	—	—	26,775	26,738	26,732	26,724
Southern Zone	7,867	12,960	14,901	18,261	23,607	33,599

Source: Pemex

period. This huge atoll-type reef extends from the Papaloapan basin 200 miles west of Reforma out to the Gulf of Campeche and encircles the peninsula offshore, crossing the coast again through Belize. It continues southwest through Guatemala and then swings north and west to close the loop back at Papaloapan.

The Papaloapan basin has yielded significant Cretaceous production, and many of the discoveries in Reforma and Campeche correlate with the highly porous calcareous pays of Papaloapan, leading to the belief that the best accumulations will be found against the barrier reef.

The Mesozoic Chiapas-Tabasco (Reforma) area

The Reforma area covers 1,916 sq miles in the states of Chiapas and Tabasco. It is bounded on the west and northwest by the Comalcalco Fault, on the east by the Frontera Fault, and on the south by the Sierra Madre de Chiapas. The discoveries in Reforma have been located primarily in Middle Cretaceous limestone and dolomites in several parallel trends running northwest-southeast. Good reservoir conditions only prevail where intense fracturing caused by uplifting has occurred.

Productive reservoirs in Reforma bear two unusual characteristics. First, they are among the rare giant oil-field reservoirs in the world deeper than 12,500 ft (12,000–14,000 ft). Second, they have uncommonly thick oil-bearing strata, ranging from 700–1,600 ft. Only a few other pay zones in the world, such as southwest Iran's Asmari limestone reservoirs, approach the maximum thickness of the Reforma fields. This thickness accounts for their relatively small ground area sizes. Per-well production (1981) is high, averaging 6,700 b/d, with wells in the Samaria field surpassing high-yield Middle Eastern fields at 16,000 b/d.

The potential of the Reforma area appears to be limited because the most promising areas are confined to the uplifted and fractured rim of the carbonate platform of the peninsula. Several exploratory wells on the flanks of productive trends have come up dry. Extension of the trend to the southwest seems unlikely given the absence of potential reservoir rock, the Cretaceous having been severely eroded.

The Campeche Sound area

The potentially productive Campeche area covers more than 3,064 sq miles in the southern part of the Gulf of Mexico. It is an apparent extension of the onshore productive trend of the Reforma area directly to the southwest. As in Reforma, the productive fields lie in parallel northwest-southeast trends in the southern part and gradually shift to a north-south alignment as they move northward. Good reservoir conditions also correspond to those of the Reforma area: the intensely fractured, uplifted perimeter of the ancient Yucatan platform.

In contrast to Reforma, the Campeche fields have been discovered at the more conventional depths of 3,500–12,000 ft. Net reservoir thicknesses are still undetermined in many cases, but indications for 2 Akal suggest a productive oil column of 8,400 ft, dwarfing the very thick Reforma structures. The *Oil & Gas Journal* gives a more conservative figure of 3,937 ft.[3] Even if that thickness proves an exaggeration, net thicknesses in the Akal block are probably twice those of the Reforma area. Due to these unprecedented dimensions and lack of oil industry experience with such thick structures, they may constitute the major uncertainty in evaluating the potential of the Campeche area.

Campeche fields seem to have other characteristics that make them even more attractive and promising than those of the Reforma area. Total porosity is better, oil in place per acre-foot (400–700 bbl) is about twice that of Reforma, and high buoyancy arising from the exceptionally thick columns has pushed productivity per well to 28,212 b/d. They are among the most prolific wells in the world (Fig. 2.1).

---

[3]For a more optimistic appraisal of the resources of Reforma-Campeche, see Bernardo Grossling, "Possible Dimensions of Mexican Petroleum," *U.S.-Mexican Relationships,* J. Ladman, D. Baldwin, and E. Bergman, eds. (1981).

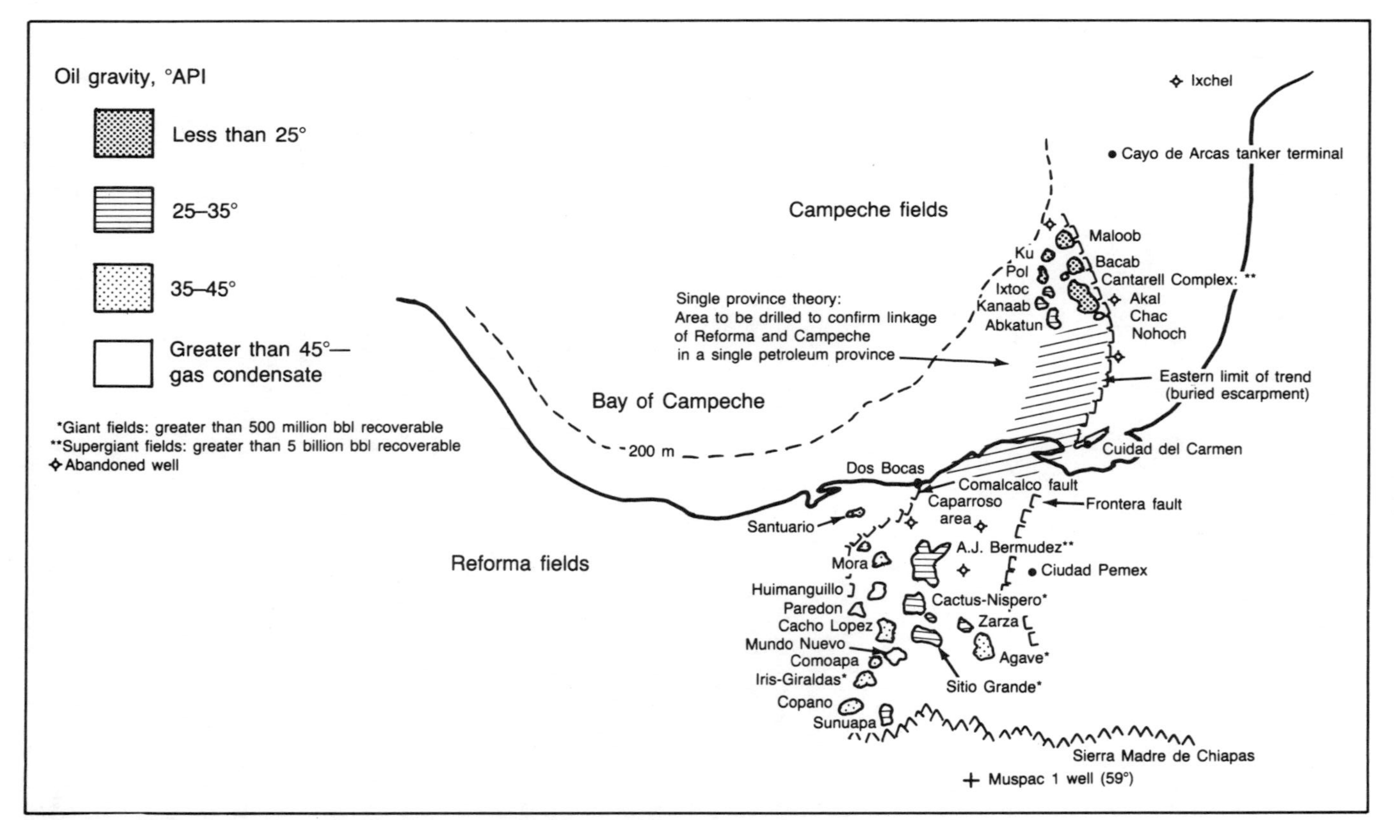

Fig. 2.1 *Oil gravity in Campeche-Reforma fields rises from NE to SW (source: Baker & Associates*

If the Reforma area and Campeche prove to be one continuous oil-bearing province, as indicated by the geologic and exploratory data acquired thus far, they may contain more crude oil than any other province in the world outside the Middle East. Pemex asserts that the results from the wells 1 Arrastradero, 1 Caparroso, 201 Cobo, and 2 Chirivital as well as the integration of geophysical information remove nearly all doubts from the proposition that Reforma and Campeche constitute a single, huge geological province. It would therefore be the only province outside Alaska and the Middle East with more than one supergiant oil field.

Through 1981, 38 exploratory wells had been completed in the Campeche area with a success rate of 82%. The area included 12 large fields. Several structures thought to be independent at first have now been linked, forming larger fields. The largest field in Campeche and in all of Mexico is the supergiant Cantarell Complex. Because production and development in Campeche have such recent beginnings, however, the ultimate potential of the offshore area still is riddled with uncertainty. The north-south boundaries of the productive trend remain obscure. Some reports indicate an absence of reservoir rock to the west and a lack of fracturing to the east that may constrict the productive trend to a strip 40 miles wide.

Prospects north and northwest of Yucatán are even less certain, since almost no exploratory drilling has been done there. Though Pemex has high hopes for the offshore Quintana Roo area, expectations arise from geologic studies and speculation, not exploratory data. Another reason for a somewhat cautious approach to the resource potential off the Yucatán peninsula is an indication, revealed by seismic work on the Ixchel structure, that absence of closure may be a general condition on the northern and eastern perimeters of the Yucatán Platform. Those areas seem to lack the overthrusting and salt movement found in the Reforma and Campeche areas. The northern and eastern perimeters also seem bereft of source rocks necessary for very large accumulations.

In spite of these concerns, a cautious 1980 study sponsored by the U.S. Department of Energy estimated that Campeche and offshore Yucatán have a 90% probability of 21 billion bbl of recoverable petroleum liquids and 12.3 trillion cu ft of natural gas; a 50% probability of 38 billion bbl, and 26 trillion cu ft; and a 10% probability of 63 billion bbl and 52 trillion cu ft.

Chicontepec

The other major contributor to Mexican hydrocarbon resources is the Chicontepec basin, which covers 4,329 sq miles in the states of Puebla and Veracruz. Total hydrocarbon resources in place are estimated to be 106 billion bbl, with 17.64 billion bbl recoverable. The accumulations in the Chicontepec area lie in sandstone beds interspersed with shale and silty shale beds. Average thickness of the productive Tertiary sandstone is 3,280 ft and reaches 6,562 ft toward the center of the field. More than 2,000 wells in the area show that the reservoir rocks have excellent correlation and very even hydrocarbon content, so the usual risks of exploration can be avoided. Pemex expects the average well in the area to produce 100 b/d initially and that drilling success will be nearly 100%. All wells in Chicontepec require fracturing before the wells will flow.

From the time the Chicontepec area was added to proved reserve figures in 1978 until the end of 1981, its contribution has remained at 17 billion bbl including natural gas, which comprises one-third of that total. The size of this resource now is well established. The issue facing Mexico and Pemex is at what rate do they wish to develop the resource? The government hopes to bring along petroleum production at the same rate as agricultural and industrial growth in the region.

The size of Mexico's commercial reserves is a controversial topic in Mexico and abroad. Jacinto Viqueira Landa writes off Chicontepec entirely, declaring that Mexico has 55 billion bbl of hydrocarbon reserves.[4] Frank Niering, meanwhile, puts oil reserves at 37 billion bbl.[5] José Luis Mejías, a Mexican journalist, attacks the credibility of Mexico's gas reserve figures.[6] The consensus is that Mexico has commercial hydrocarbon deposits in the range of 30–40 billion bbl of oil and 10–15 billion bbl equivalent of gas.

Return to World Stature

To appreciate the magnitude of Mexican oil efforts during the period 1977–81, some comparative perspective is instructive. Contrary to forecasts at the beginning of the period, annual average world production had stabilized and OPEC production had shown a decline of over 27%. Communist crude production was 23 million b/d, an increase of 9% from the annual average of 1977—most of this growth, 1.2 million b/d, having taken place in the Soviet Union. But the largest growth in crude production, more than 45%, took place among non-Communist LDCs outside of OPEC.

In 1977 the U.S. produced nearly 14% of world output, more than one-fourth of the volume produced by OPEC and more than twice the volume of non-OPEC LDC producers. Two OPEC member nations, Venezuela and Ecuador, together produced more than 4% of world production, nearly 8% of OPEC production, and the equivalent of nearly 60% of the total volume of non-OPEC LDC production. Canada's production, more than 30% greater than that of Mexico, was equivalent to 30% of the total production of non-OPEC LDCs. Mexico in 1977 was responsible for less than 2% of world production but for nearly one-fourth of the production of non-OPEC LDCs.

---

[4]Professor at Universidad Nacionale Autónoma de México.

[5]*Petroleum Economist* (July 1982), p. 277.

[6]*Excelsior* (April 4, 1983), p. 1.

In 1981, U.S. production was up less than 4% from 1977. The combined production of Venezuela and Ecuador was now the equivalent of only 39% of non-OPEC LDC output. Canada's production was almost one-half less than that of Mexico whose crude volume, as a proportion of non-OPEC output, had grown nearly 63% since 1977.

In 4 years, Venezuelan crude production had dropped 6%, while that of the U.S. had increased by nearly 4%. Canadian production was down more than 2%; Ecuador's was up 15%. Mexican production growth had far outpaced other Western Hemisphere producers, up nearly 136%.

In 1981 the relative contributions of the U.S., Canada, Ecuador, Venezuela, and Mexico had increased. Mexico nearly doubled its contribution to world supply to more than 4%. Each of these five Western Hemisphere producers increased its crude output relative to the total production of OPEC. Mexico's relative growth far exceeded that of any of the others. Mexico's production was now the equivalent of more than 10% of that of OPEC. In 1977 Mexico produced not quite a quarter of non-OPEC LDC oil, but in 1980 Mexico's share had grown to 39% (Table 2.4).

Table 2.4
The rise of Mexico's oil stature

	1977	*1978*	*1979*	*1980*	*1981*
Mexico oil production*	981	1,209	1,461	1,937	2,313
Mexico/World (%)	1.64	2.01	2.33	3.26	4.15
Mexico/OPEC (%)	3.14	4.06	4.72	7.20	10.20
Mexico/non-OPEC LDCs (%)	23.99	27.29	29.51	35.50	39.00

*Thousand b/d, excluding NGL
Source: Central Intelligence Agency

With the settling in of oil surplus conditions in 1981–82, Mexico was forced to conclude that all oil exporters, OPEC and non-OPEC nations alike, shared a common market destiny.

Oil in place is an abstraction. For Mexico during López Portillo's time, it stood as the foundation of Mexico's line of credit with international bankers. Turning Mexican oil abstractions into daily production of crude, gas, and product would be expensive, especially once the costs of imports and inflation, at home and abroad, were taken into account.

3

Government investments in the oil sector

Because all other petroleum investments are tied to capital spending for production work, a study of Mexican government investments in the oil sector must begin with a discussion of investments in production. This category will illustrate the cost of investments in the industry to the Mexican economy. The methodology applies to each category of Pemex's investments.

Pemex invested more than originally planned in every category. Total planned investment for the 5-year period was $12.717 billion while actual investments reached $20.048 billion. That $7.3 billion difference represented a very large underestimation by Mexican planners of the investments needed to reach the production targets. Since in most categories the physical plant to be created by the investments did not reach levels programmed, the planners also overestimated the purchasing power of their pesos.

Pemex Investments in Hydrocarbon Production

The calculation of Mexico's investment costs for crude oil and natural gas provides the basis for understanding the power and limitations of the oil sector to respond to Mexico's liquidity crisis that began during 1981–82. That raises several questions: What is the average investment cost per barrel that Mexico incurred during the López Portillo era? Moreover, what were these costs onshore vs offshore, as a result of the development of the Campeche fields? Another question concerns the marginal costs per barrel onshore and offshore. While the average cost per barrel is a measure of the costs during a period, the marginal cost per barrel tells the cost of producing the next barrel at a particular point on the supply curve.

Classical economic theory teaches that unit costs at first decline and then ultimately rise. This is another way of saying that as greater efficiencies are achieved, greater returns to scale (lower unit costs) occur. At some point declining costs reach their minimal value and then begin to rise. This process happens for both average and marginal costs.

A method for estimating average investment costs per barrel is needed to determine cause-and-effect relationships. The magnitude of effect (measured as new, or net, oil or gas production) should be proportionate to the magnitude of effort measured as investment costs in constant units, netted of distortions such as inflation. Thus, the first two analytical tasks are to convert investment and output information into a useful form. The result would be that the investment of a period is matched against its contribution to new output.

The first, easier task is to convert investments by Pemex in hydrocarbon production (reported in current dollars to bankers) to constant units, taking into account the fact that some of those "dollars" were spent in Mexico as pesos. To create constant equivalent dollars, Mexican inflation on peso expenditures and U.S. inflation on dollar expenditures would have to be washed out.

One argument is that since the peso was overvalued during most of the López Portillo period, Pemex was in effect subsidized in its imports of goods and services by a factor equal to the degree to which the peso was overvalued. Investment costs to the Mexican economy are therefore higher than those investment costs as reported by Pemex. Pemex, one can argue, is an accounting fiction from the standpoint of national accounts.

During 1976–80 the smell of oil caused international lenders to lend to Pemex, knowing that a good part of the money went to the Mexican government for general revenue purposes through a 90–99% tax on operating profits. It did not matter who got the money as long as that someone, through Pemex oil sales, could repay it. This consideration leads to a distinction between *Pemex dollars* of a given year and *parity exchange rate dollars* (or parity dollars, for short) for the same year.

The second task is to determine *net* output from *gross* output in a given year. To know that Pemex produced so many barrels more in 1981 than in 1980 does not, for several reasons, tell you what investment caused that increase. You have to know the time lag between investment and output. In Mexico's recent experience this has been between 6 months and 2 years. Further, if 100 b/d were produced in 1980 and the depletion rate was 10%, then 90 b/d were flowing in 1981. If total average output in 1981 was 200 b/d, then the increase to 1981 from 1980 was 110 b/d, not 100 b/d. If the time lag between investments and output was less than one year, those 110 b/d were "caused" by the corresponding investment for 1981. But one must first convert Pemex dollars to fully loaded costs to the Mexican economy.

Pemex production investment costs in 1981 constant dollars

Typically, the amount of the capital expenditures by developing countries on crude production is not readily available outside those countries. Some figures are circulated on a strictly confidential basis to prospective institutional lenders who require, if only as a matter of formality, abundant financial and industrial statistics. To an audience of English-reading bankers, it makes sense to translate foreign currency into current dollars at official exchange rates. Such a procedure allows participating banks to make their own evaluation and adjustment, if any, for an overvalued currency. The effect is to understate the cost to the total economy of dollars purchased at subsidized rates.

The problem of how to establish a basis for evaluating crude production costs to the economies of countries such as Mexico is perplexing. The state oil company *a*) sells oil in dollars, *b*) receives payment in local currency (via the country's central bank), and *c*) buys dollars for expenses and capital imports at subsidized exchange rates.

The figures for Pemex's capital expenditure circulated in the U.S.—they are not published in Mexico—are those given to Pemex's bankers and are expressed in current dollars. Not that Pemex in a given year actually spent that many dollars, but that Pemex's expenditures, originally expressed in pesos, divided by official exchange rates produce dollar-like numbers that meet the needs and concerns of lending institutions.

Such dollar figures can be misleading. In the first place, there is an inherent ambiguity in what the term "capital expenditures" might mean. Does it mean the sum of the pesos spent during the year in the given area of activity or the value of the new capitalized assets, tangible and intangible, acquired during the year and remaining on the books on December 31? If amounts given represent money spent uniformly during the year, then to express this value in dollars the average exchange rate for the year would apply. If the amounts represent asset values on December 31, then the year-end exchange rate would apply.

This question becomes especially important in a year where there is a devaluation of the currency against the dollar. Consider Pemex's reported expenditures for development and production in 1976. The total value reported is $448 million translated at 20 pesos to the dollar. If the sum of Pemex's new year-end assets in these two categories was 8.954 billion pesos, then the figure of $448 million would represent its dollar-equivalent worth on December 31.

On the other hand, if those 9 billion pesos were spent throughout the year, then the year's average exchange rate, 15.44:1, would apply; in that case, Pemex's equivalent dollar expense would be $580 million, not $448 million. Pemex's development and production expenditures in 1976, if converted into pesos, would represent an amount more than twice the corresponding amount in 1975; but if left in dollars at the year-end rate, 1976's figure would represent a much more believable increment of 36%.

In either event, Pemex spent some amount of pesos in 1976 to buy dollars at an exchange rate of 12.5:1 from January through August and ended up buying at 19.95:1 on December 31.

Total investment in production is taken as the sum of development drilling and production investments. Pemex's bankers give figures for these categories for the period 1974–1981 (Table 3.1). To make these data useful, some accounting adjustments must be made for the effects of an overvalued peso and inflation, domestic and foreign, during the 8 years.

Table 3.1
What Pemex spends for drilling and production

	(Millions of current dollars)							
	1974	*1975*	*1976*	*1977*	*1978*	*1979*	*1980*	*1981*
Development	149	184	191	176	289	379	683	845
Production	118	145	256	280	831	1,223	1,472	1,904
Total	267	329	448	455	1,120	1,602	2,155	2,749

	*(Millions of current pesos)**							
	1974	*1975*	*1976*	*1977*	*1978*	*1979*	*1980*	*1981*
Development	1,857	2,298	3,828	4,039	6,647	8,724	15,698	20,354
Production	1,476	1,810	5,126	6,435	19,121	28,128	33,860	45,836
Total	3,334	4,108	8,954	10,474	25,768	36,852	49,557	66,190

*Unrounded dollar figure times the average annual exchange rate.
Source: Pemex

The first step in adjusting for the overvalued peso is to convert current dollars to pesos at the average official exchange rate for each year.

Second, since capital expenditures by Pemex are made in both pesos and dollars, only that portion of expenditures that went to buy dollars should be adjusted for the impact of an overvalued peso. One-half of the capital expenditures is assumed to be in dollars, based on private industry estimates ranging from 40–80%.

For the sake of argument, a conservative estimate of 50% of peso expenditures is taken as the nominal peso cost for the dollars used to purchase foreign capital goods. The nominal cost is divided by the average official exchange rate to get the dollars received by Pemex (Table 3.2). If these dollars are converted back to pesos at an exchange rate adjusted for differences of inflation in the U.S. and Mexican economies, the real peso cost of those dollars is arrived at.[1]

Subtracting Pemex's peso cost from the real peso cost tells how much Pemex's purchase of dollars was subsidized by the overvalued official exchange

[1]To arrive at a parity-adjusted exchange rate for each year, take the 1960 exchange rate, 12.5:1, multiplied by the value of the year's GPD deflator as the base year, and divide the product by the U.S. deflator having the same base year.

Table 3.2
A look at real cost to Mexico of Pemex's D & P

	(Millions of current parity-adjusted pesos)							
	1974	*1975*	*1976*	*1977*	*1978*	*1979*	*1980*	*1981*
Official exchange rate	12.50	12.50	15.44	22.58	22.77	22.81	22.95	24.51
Parity rate	16.21	17.31	20.03	24.99	27.51	30.61	36.15	41.93
Development								
Local currency	929	1,149	1,914	2,019	3,324	4,362	7,849	10,177
Dollar expense								
Pemex's cost	929	1,149	1,914	2,019	3,324	4,362	7,849	10,177
Ex-rate subsidy	276	442	569	216	692	1,492	4,514	7,233
Real peso cost	1,204	1,591	2,483	2,235	4,016	5,854	12,363	17,410
Total	2,133	2,740	4,397	4,254	7,339	10,216	20,212	27,587
Production								
Local currency	738	905	2,563	3,218	9,560	14,064	16,930	22,918
Dollar expense								
Pemex's cost	738	905	2,563	3,218	9,560	14,064	16,930	22,918
Ex-rate subsidy	219	348	762	343	1,990	4,809	9,738	16,289
Real peso cost	957	1,253	3,325	3,561	11,551	18,873	26,667	39,207
Total	1,695	2,158	5,888	6,779	21,111	32,937	43,597	62,125
Total D & P	3,828	4,899	10,285	11,033	28,450	43,153	63,809	89,712

Source Pemex; adjustments by Baker & Associates

rate in a given year. Parity-adjusted investments in hydrocarbon production, then, equal the local capital expenditures on development drilling and production plus the real peso cost of dollar-denominated expenditures associated with the foreign-exchange component of the capital investment.

These parity-adjusted figures represent an indirect subsidy benefitting Pemex and a cost to the Mexican economy. The subsidy to Pemex is equal to the difference between Pemex's costs of investing at the official rate of exchange on its dollar transactions and costs that Pemex would have incurred in an open-currency market. The most obvious cost to the economy is in the dampening effect of an overvalued peso on Mexican exports—including tourism, until recently Mexico's largest source of foreign exchange—with the corresponding encouragement of imports and the consequent negative effects on Mexico's balance of payments and foreign reserves.

Next, inflation is considered as a factor in the true investment costs (as distinct from operations, financial and depreciation costs) of producing hydrocarbons in Mexico (Table 3.3). The first step is to take local current

Table 3.3
How to wash inflation from Pemex's D & P investments

	(Millions of parity-adjusted 1981 pesos)							
	1974	*1975*	*1976*	*1977*	*1978*	*1979*	*1980*	*1981*
Petroleum & petrochemical deflators								
1960 = 1.0	1.505	1.722	1.862	2.755	3.113	4.296	5.80	6.65
1981 = 1.0	0.23	0.26	.28	0.41	0.47	0.65	0.87	1.00
Dollar import deflators								
1960 = 1.0	1.908	2.115	2.174	2.286	2.485	2.776	3.109	3.27
1981 = 1.0	0.58	0.65	0.66	0.70	0.76	0.85	0.95	1.00
GDP deflator								
1960 = 1.0	2.17	2.532	3.081	4.069	4.807	5.802	7.466	9.459
1981 = 1.0	0.23	0.27	0.33	0.43	0.51	0.61	0.79	1.00
Development								
Local currency	4,103	4,438	6,836	4,874	7,100	6,753	8,999	10,177
Dollar expense								
Pemex's cost	1,592	1,777	2,879	2,889	4,374	5,139	8,255	10,177
Ex-rate subsidy	1,201	1,652	1,747	501	1,361	2,432	5,719	7,233
Real peso cost	2,793	3,429	4,626	3,390	5,735	7,570	13,975	17,410
Total	6,896	7,866	11,462	8,264	12,835	14,323	22,974	27,587
Production								
Local currency	3,261	3,495	9,153	7,767	20,423	21,770	19,411	22,918
Dollar expense								
Pemex's cost	1,265	1,399	3,855	4,603	12,581	16,567	17,807	22,918
Ex-rate subsidy	955	1,301	2,339	798	3,916	7,840	12,337	16,289
Real peso cost	2,220	2,700	6,194	5,401	16,497	24,407	30,144	39,207
Total	5,481	6,195	15,347	13,168	36,920	46,177	49,555	62,125
Total D & P	12,378	14,062	26,809	21,432	49,755	60,500	72,528	89,712

Source: Wharton/Diemex; Pemex, with adjustments by Baker & Associates

expenditures (50% of the total) and divide them by the deflator for the Mexican petroleum and petrochemical sector to get local expenditures in constant 1981 pesos.

Second, the effect of inflation on dollar-denominated expenditures has to be calculated. Since those goods are valued in dollars, a different deflator applies. One-half of the total current peso expenditures, representing the nominal cost of dollars, is divided by the official exchange rate to get actual dollars received.

Those dollars are then divided by the dollar import deflator to express dollars in 1981 values, thereby accounting for U.S. inflation as it affects Pemex purchases in dollars. The 1981 dollars multiplied by the official exchange rate give the 1981 peso equivalent of those dollars adjusted for U.S. inflation. Thus far, the effects of Mexican inflation in the petroleum and petrochemical sector have been taken into account as they affect local Pemex capital expenditures and the effects of U.S. inflation as it pertains to Pemex capital expenditures in dollars.

The third step is to recalculate the value of the exchange-rate subsidy to Pemex in 1981 pesos. As this subsidy is absorbed by the Mexican economy in general, the GDP deflator is used, yielding the calculated values of this subsidy for development and production expenditures.

The total investment cost for development and production expressed in parity-adjusted 1981 pesos is converted to millions of 1981 constant equivalent dollars to represent the cost to the total economy of Pemex's investment in new and replacement production (Table 3.4). As expected, these numbers are higher than nominal numbers owing to the adjustments made for *a*) inflation in Mexico's own petroleum industry, *b*) inflation in the cost of imported goods and services (probably understated), and *c*) changes in the general price level, taken as the measure of the rising cost to the economy of maintaining exchange-rate subsidies. Adjusted figures for 1981 are higher than the nominal figures for 1981 by the dollar-equivalent value of the exchange-rate subsidy.

These adjusted figures become, then, the fully loaded amounts (exclusive of capitalized interest payments) invested in hydrocarbon production.

The ratio to be used for converting nominal investment figures to "parity" dollars can be calculated by dividing 1981 parity dollars by nominal dollars (Table 3.5).

One can argue that this procedure applies where inflation in the economies of oil-exporting countries is substantially higher than that in the U.S., where exchange-rate policies have kept an artificially rigid relationship between local currency and the dollar and where dollar-imports of goods and services account for a substantial portion of development and production costs.

Venezuela falls under this category. There, the government makes an admirable point of calculating national profits per barrel, including, by that

Table 3.4
Pemex's true D & P investments

	(Millions of 1981 constant equivalent dollars)							
	1974	*1975*	*1976*	*1977*	*1978*	*1979*	*1980*	*1981*
Development	281	321	468	337	524	584	937	1,126
Production	224	253	626	537	1,506	1,884	2,022	2,535
Total	505	574	1,094	874	2,030	2,468	2,959	3,660

Source: Tables 3.2 and 3.3

Table 3.5
Factors to convert Pemex-reported current dollars for D & P to 1981 parity-adjusted constant dollars

	1974	*1975*	*1976*	*1977*	*1978*	*1979*	*1980*	*1981*
Total D & P								
Table 3.1 (A)	267	329	448	455	1,120	1,602	2,155	2,749
Table 3.4 (B)	505	574	1,094	874	2,030	2,468	2,959	3,660
Conversion factor (B/A)	1.89	1.75	2.44	1.92	1.81	1.54	1.37	1.33

term, taxes paid by oil industry contractors and suppliers distributed over annual crude production. In general, the state oil company Petroleos de Venezuela (PDVSA) was light-years ahead of Pemex in the area of financial disclosure and reporting, as anyone knows who has ever looked at the skimpy financial section of Pemex's annual reports.[2]

Despite this more open policy in the area of statistical reporting, in this case Venezuela's oil ministry understates PDVSA's costs and overstates national profits by the effect of the implicit subsidy whose value is represented by the difference between the value of dollar imports calculated at the official exchange rate for the bolivar (long at 4.30:1) and the value of the same imports calculated at a parity-exchange rate that cancels the inflation differential between the economies of Venezuela and her trading partners.

Of course neither PDVSA nor Pemex can report on topics about which their respective governments choose to remain silent. In Pemex's case, the Mexican government has refused to discuss 1982's collapse of the value of the peso against the dollar as the outcome of a deliberate policy of maintaining an overvalued peso. Instead, it has blamed its banking system for selling its dollars to unpatriotic Mexicans. The basic truth, however, is that the government, through its exchange-rate policy, had been subsidizing the purchase of dollars by Pemex and other importers and capital exporters. Pemex, in turn, as a patriotic service organization, is in no position to adjust upward the value of goods and services purchased abroad. On the contrary, through the vehicle of letting contracts with the Mexican Petroleum Institute (IMP), Pemex is able to pay in pesos "to a Mexican supplier" for goods and services that the supplier in turn may have to import. In countries like Mexico, Venezuela, and Indonesia where local procurement is an ideological imperative, such cost accounting niceties are likely to be left to the foreign analyst.

Exploration Investments

Between 1977 and 1981, the Pemex program designed in the first year of the López Portillo administration called for investment in exploration to total

[2]The two volumes reporting Pemex's financial condition in 1982 represent a profound departure from previous practices.

$745 million. Actual investment, in dollars calculated with the official exchange rate, totaled $919 million or a 23% jump from the programmed total.

That's a far cry from the earliest days of Mexico's oil sector. When President Lázaro Cárdenas signed the bill of expropriation on 18 March 1938, geological information recovered from foreign petroleum companies was as rare as foreign support for the move. From that time forward, the neophyte Pemex began the slow and tedious process of assessing the nation's sedimentary basins.

Pemex's long apprenticeship in the exploration business finally bore very rich rewards during the 1970s. The oil price increase of 1973, with attendant financial problems for the oil importers, had Mexico along with much of the world scrambling for relief. By the time the López Portillo administration entered the scene in 1976, accumulated geological information allied with the impetus of high oil prices to inspire new exploratory activity.

Seismic and particularly geological surveying increased at rapid rates beginning with the López Portillo administration (Fig. 3.1). At the same time, the crew-months put in by Pemex's interpretation and evaluation groups went up at a similarly steep rate, increasing by 64.2% from 1977 to 1978 alone. This

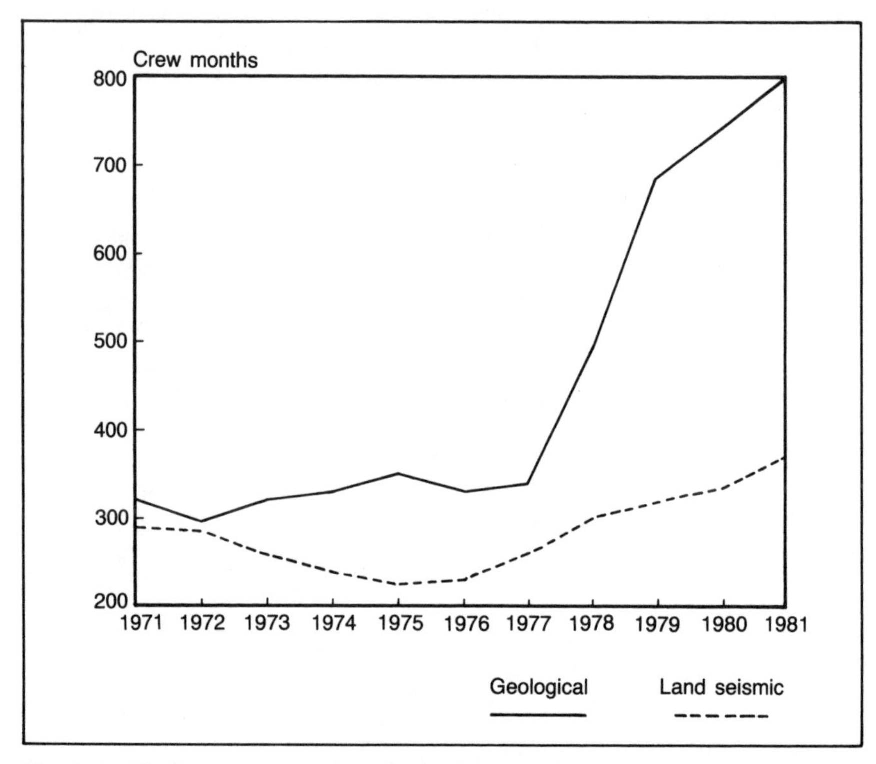

Fig. 3.1 *Exploratory surveying climbs (source: Pemex)*

stepped-up exploratory activity and analysis translated into dramatic additions to Mexico's proved reserve figures.

During the 1970s, only three supergiant fields (in excess of 5 billion bbl recoverable) were discovered in the world; all three are in Mexico. Those and other discoveries meant that Pemex added more oil and gas to proved world reserves than any other company, public or private, in that decade. In 1976, proved reserves were at 11.1 billion bbl of oil equivalent, jumping to 40 billion bbl by 1978, and were at 72 billion bbl by the end of 1981.

Mexico's "priority areas," areas with the greatest potential to add quickly to reserves, are the Mesozoic Chiapas-Tabasco region (Reforma) and the Gulf of Campeche. The Reforma area was discovered in 1972 by the 1 Cactus and 1 Sitio Grande wells. Its 2,703 sq miles have yielded 30 producing fields through 1981. Campeche made its abrupt appearance in 1976 through the 1 Chac well located 50 miles offshore. The 3,088 sq miles offshore hold 12 known large fields.

Other areas have received attention, too. In the states of Coahuila and Nuevo Leon lies the 15,440-sq-mile Sabinas basin, site of eight structures currently producing exceptionally clean natural gas. Many similar structures have been identified. About 75% of total structures comprise still-virgin territory. Though the Sabinas basin has always been considered an exclusively gas-producing zone, in 1981 Pemex found oil at 1 Vivianco near Monclova. A mere 10.5 miles off the mouth of the Colorado River lies the gas well 1 Extremeno, the first commercial discovery on Mexico's west coast. That encouraging find, made in 1981, will support further exploration in the area, which had been waning.

Before 1976, only 10% of Mexico's surface area had been exhaustively evaluated for its hydrocarbon potential. Another 5% was covered by 1981. Pemex claims that of Mexico's 960,000 sq miles, including the continental shelf out to a 1,640-ft depth, only 20% has no petroleum potential.

Government Drilling Investments

The Pemex program did not segregate exploration and development drilling, only giving figures for total drilling, at $3.202 billion the largest category of investment in the entire program. Actual investment for total drilling from 1977 through 1981 was $4.555 billion, a 42% increase from the figure planned at the beginning of the López Portillo administration.

Actual wells drilled, both exploratory and development, fell far short of the targets set in the same plan that projected capital investments. The planners clearly underestimated the investment levels required to sink the number of wells that were considered necessary to reach desired production targets. Fortunately for the industry, the fields discovered and the wells completed proved more prolific than anticipated by the administration's planners so that production targets were still reached—and ahead of schedule for both crude oil and natural gas (*see* Appendix E).

Under López Portillo, Pemex's exploration program was to extend to 28 of the 31 states of the country in addition to the Bay of Campeche in the Gulf of Mexico and to the Pacific Ocean. Pemex expected to drill 1,324 exploration wells betwen 1977 and 1982 or more than twice the number of wells drilled during the previous 6-year period. Only 641 exploratory wells had been drilled during the previous 6 years, of which 143 (22%) resulted in successful oil and gas strikes.

Since the beginning of the program, Pemex has added more than 60 billion bbl of hydrocarbons to Mexico's proved reserves, or increases of about 10 billion bbl per year. One might expect such a meteoric rise to be reflected in rapidly expanding figures for exploratory drilling along the lines of the Pemex program. In fact, only 418 wells vs a targeted 1,324 were drilled. The number of drilling rigs operated by Pemex and devoted to exploration varied between 53 and 70. The footage actually drilled rose slightly through 1978, but then fell steadily through the remainder of the period. Exploratory wells completed during those 6 years follow a somewhat bell-shaped growth curve, starting at 79 wells, going up to a high of 85, then dropping to 70 in 1981.

That year also marked the first year since 1975 that Pemex went without a discovery in the Gulf of Campeche. A need for more light oil during the year led to Pemex's decision to divert exploratory rigs from Campeche to development drilling. Not until the final quarter of the year were three rigs again assigned to exploratory work in the Gulf.

The surprising fact is that one has to go back to the early 1950s to find a succession of years with as few exploratory wells sunk per year as during 1976–81.

So why did reserves increase dramatically during the López Portillo administration while exploratory drilling demonstrated a declining trend? Part of the answer lies within the mysterious realm of the proved reserves calculus. As the variables of world oil price, technology, and geological insight change, so do proved reserve figures, even without added discoveries of new fields.

As Jorge Díaz Serrano, the former president of Pemex, explained, the company reevaluated its definition of proved reserves at the beginning of 1977, believing the calculus had erred in the past on the side of excessive conservatism. He said it had been done "without taking into account that it is just as serious to err with low figures, as with high ones, and that incorrect conservative estimates cost enormous sums of money because they lead to planning at inefficiently low levels."

The rest of the answer comes from the new drilling strategy adopted during the period. For most of Mexico's and Pemex's history, production in the states of Chiapas, Tabasco, and Veracruz was concentrated in Tertiary geological regions. During the 1970s, wells were sunk to the much deeper Cretaceous and Jurassic formations where great riches were tapped with relatively few wells.

The 480 exploratory wells drilled between 1976 and 1981 had a very respectable success rate of 36.6% (the U.S. was less than 30% for the period). However, the two highest priority exploration areas, Reforma and Campeche, had enviable success rates of 52% and 82%, respectively. Those impressive figures point to two prominent features of exploration during the 1976–81 period.

First, Pemex concentrated on those two areas as offering the greatest potential for adding quickly to reserve figures. Pemex stuck with winning areas, not squandering resources on a lot of wildcatting in less-promising regions. Second, given the overall success rate of 38.2%, the rate had to be very much lower for areas outside the known prolific areas of Reforma and Campeche.

In fact, drilling on the western side of Baja California was suspended in 1981, and the drillship was reassigned to Campeche. Three dry holes in the Salina Cruz geological province on the continental platform of the Pacific convinced Pemex to end drilling in that area for the indefinite future. Indications from the few wells drilled in these areas are that the rocks are geologically too young to have formed accumulated hydrocarbons.

Offshore Mazatlan has a promising 19,680-ft thick sedimentary column, but the structures are soft, making it difficult to establish closure. Two dry holes and one noncommercial discovery were drilled. Three more exploratory wells are planned for the area. Pemex classifies Mazatlán offshore as a potential petroleum province. With the exception of the 1 Extremeno gas well in the Sea of Cortés, drilling offshore in the Pacific was unproductive for Pemex during the 6-year period.

Development drilling

Pemex announced in December 1976 that it expected to drill 2,152 development wells between 1977 and 1982. About 475 of the new development wells were expected to be drilled in the states of Chiapas and Tabasco and 120 on the continental shelf in the Bay of Campeche. In contrast to the downward trend in exploratory drilling evident in the second half of the 1976–81 period, development drilling did pick up. Actual drilling footage demonstrated the opposite trends in the two drilling categories: as development footage increased, exploration footage dropped (Fig. 3.2). Acting like opposing magnetic fields, the trend lines illustrate the dilemma facing Pemex due to the inverse relationship between development drilling and exploratory drilling.

Despite a 9.4% average annual increase in the number of drilling rigs in operation during the 6 years, there were not enough rigs to meet the demand for increased production, which came primarily through the development of new fields in Reforma and Campeche. Given that requirement, Pemex was forced to

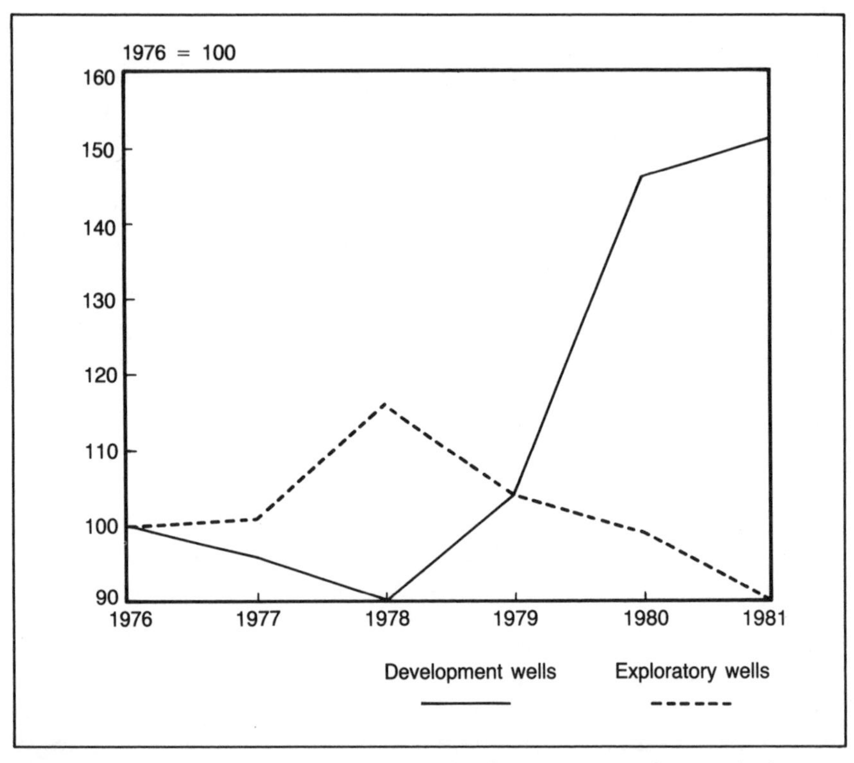

Fig. 3.2 *Inverse drilling relationship (development vs exploratory), (source: Baker & Associates)*

withdraw rigs from exploration and put them to work on developing new finds. Not once in the 6 years did both types of drilling activity increase in the same year.

In 1981, the Chiapas-Tabasco Mesozoic region was home to 36% of the Pemex rig fleet. Development wells were drilled in 70 fields—43 oil and 27 gas; 66 onshore and 4 offshore. In the Campeche Sound, Pemex had 14 platform rigs and 2 jackups working, while 8 platforms were in place awaiting rigs and 12 other platforms were under construction or being installed.

This accomplishment comes very close to the development schedule for the Campeche fields set up shortly after their discovery. Pemex's production plans called for completion of 120 wells on 20 platforms by the end of 1982. With 14 platforms working and 8 more awaiting rigs in 1981, Pemex was within reach of the 20-platform rig target.

The only significant gap between planned and actual performance concerns the number of wells. Aiming for 120 wells by 1982, only 55 wells were producing by the end of 1981, too low for the difference to be made up in 1982.

As in the case of overall development drilling, the very high productivity of each well in offshore Campeche obviated the need for as many wells as were originally planned. Had the wells been of a more average flow rate, Pemex would have fallen far short of both scheduled well completions and production targets.

The biggest jump in development drilling came from 1979 to 1980 when the number of wells went to 349 from 250, a 40% increase. Although the number of wells dropped in 1981, footage drilled still increased from 1980's total. Not only did drilling activity increase, but its efficiency also was on the rise. From a success rate of 78% in 1976, Pemex's development holes hit a gratifying rate of 87% in 1981, with drilling in Campeche hitting 100% that year. These high success rates, coupled with prolific production per well, account for Pemex reaching its production targets despite sinking only 1,581 wells where the planned level was 2,152 for 6 years.

Pemex did most of its own drilling during the 1976–81 period. Contractors completed less than 5% of total development footage through 1978 when they drilled only 1.8% of footage. Since then, contractors have picked up a growing share, reaching 10% in 1979, 24.2% in 1980, and 30.8% in 1981. In exploratory drilling, contractors started to gain a larger share of footage drilled, reaching 41% in 1979, but then falling to 9.8% in 1981.

Mexican rig productivity is low by U.S. standards, sinking only two wells and 20,000 ft per year. This abysmal level is due to the depth that each borehole must plumb to hit pay dirt. The average Mexican well is more than 10,000 ft deep. U.S. wells are less than half as deep, and productivity per rig is 20 wells and 100,000 ft per year.

Shortly after the 1973 oil crisis, President Luis Echeverría Alvarez called together a group of 100 top Mexican industrialists and urged them to develop the capital goods needed to exploit Mexico's oil wealth. The intent was to avoid dependency on foreign sources of technology and capital goods such as drilling rigs. Even as early as that meeting, Pemex realized that much of Mexico's oil potential lay at very great depths, and its exploitation would put a severe strain on its fleet of drilling rigs. That strain clearly was exacerbated as exploration and development moved into deeper pay zones and offshore areas.

However, President Echeverría's exhortations and the obvious market potential did spur Mexican industry into action. Industria del Hierro and Conjunto Manufacturero of the Lanzagorta group, two Mexican engineering firms, delivered the first land drilling rigs to Pemex in late 1976. The 159 Samaria well in Reforma hit a pay zone on 30 March 1977 and became the first productive well in Mexico sunk with Mexican-made equipment.

Another step toward self-sufficiency in drilling rigs was made in 1981 when the first offshore oil platform ever built in Latin America began drilling in Campeche. The PM4054 marine rig was built in less than a year by Landermott

S.A. de C.V., an affiliate of the Lanzagorta group. The new rig is in operation on the 74 Abakatun shelf and is the first in a series of 16 being built for Pemex. Each rig is capable of drilling 12 directional wells in each location and can be moved to another platform and drill in a new position. On this first rig, 43% of the components came from Mexican manufacturers, and this proportion will be increased to nearly 70% in later versions. The quality of Mexican-made rigs has attracted attention, and orders have been placed from Brazil and the U.S.

Petroleum Processing Investments

The Pemex plan for the years 1977–81 called for investing $2.041 billion in refining, approximately one-half billion less than actual investments, which were $2.619 billion.

The Pemex refining expansion program included plans for five major new refineries by 1982 and a doubling of capacity from the 1976 level. By the end of 1982, then, capacity was to have reached 1.67 million b/d. That goal was not met but did increase from 785,000 b/d in early 1976 to 1,620,500 b/d by the end of 1982, a jump of 106%. Four new refineries came onstream between 1977 and 1981.

Programmed investments in basic petrochemicals were $2.283 billion from 1977–81. Actual investments exceeded that figure by over $1 billion, a 45.7% increment over the planned level of investment.

Investments in Transportation and Distribution

The López Portillo plans for investment in this area foresaw a total expenditure of $1.687 billion from 1977 through 1981. The actual level was only a bit higher at $1.852 billion. Significant advances were made in improving the infrastructure in this area.

With 6,200 miles of coastline and industrial ports facing both Atlantic and Pacific markets, Pemex, no less than the national economy at large, has long-range incentives to develop its maritime transport capabilities.

Maritime investments

In 1935, Pemex and maritime transport were nearly strangers. Total capacity at that time was 34,000 deadweight tons (dwt), with an average tanker age of 25 years. Between 1976 and 1981, 12 new tankers with a combined capacity of 473,900 dwt were added to the Pemex fleet, or more than one-half of the fleet total of 900,374 dwt at the end of 1981. Crude capacity was 6,789,964 bbl. Average fleet age for the 35 vessels was 9 years and 67% of the total tonnage was in ships under 9 years of age.

During the period 1976–81, Pemex had to rent vessels to supplement its own fleet. For example, during the first quarter of 1979, the Pemex fleet transported 14.4 million bbl of crude products and 600,000 bbl of petrochemicals—levels 15% and 40%, respectively, lower than the programmed goals.

According to Pemex management, these shortfalls and similar problems in other periods were due to the lack of storage facilities at some terminals. The rented fleet transported 4.8 million bbl during this representative quarter, 81% of its crude oil. In addition, buyers' ships handled 29.8 million bbl, mostly crude for export. The general trend during the year was for the Pemex fleet to handle 30% of total maritime shipments, the rental fleet 10%, and buyers' ships the remaining 60%.

In order to supply growing internal demands and meet export levels, Pemex still had to rent a number of vessels in 1981: 61 for the movements of the ship, pipe, and port project; 31 auxiliary units for port and fluvial operations; and 43 ships used in the general cargo system.

Mexico's major oil ports are at Pajaritos, Tuxpan, and Ciudad Madero on the Gulf Coast and Salina Cruz and Rosarito on the Pacific. Ships of 45,000–155,000 dwt can be accommodated at these ports. In 1979, construction began on another terminal with the ability to handle tankers of 250,000 dwt at Dos Bocas, just north of Villahermosa. Another 18 minor ports handle vessels of 1,000–3,000 dwt, mostly for the transport of refined products (Fig. 3.3).

Pajaritos, the site of the Minatitlan refinery, is the most important crude oil export terminal in Mexico. In 1981, it had eight berths with 43-ft depths, each capable of loading 55,000-dwt vessels, and one offshore buoy in 85 ft that could handle tankers up to 150,000 dwt. In 1980 and 1981, six more berths were put under construction, and another offshore buoy, planned for installation in 1980, was delayed. At the end of 1981, Pajaritos had an export capacity of about 1 million b/d and storage capacity of 3 million bbl with 2 million bbl more being added.

The capacity at Pajaritos has been supplemented by a stationary oil tanker anchored in the Sound of Campeche off Cayo Arcas with a capacity of 200,000 b/d. Oil is transferred from it to tankers for storage in Curacao and then transshipment to European destinations.

Dos Bocas is scheduled to become Mexico's main crude oil export terminal. Shortly after the potential of the offshore Campeche fields became apparent, Pemex selected Dos Bocas as the site for both the major export terminal and the base from which to supply all of the support services necessary for the development and operation of the offshore fields in Campeche Sound. Two 36-in. crude oil pipelines from offshore Campeche reach land at Dos Bocas, one of them going on to Pajaritos. By the end of 1981, Dos Bocas had a storage capacity of 2.4 million bbl and a loading buoy 12.4 miles offshore capable of handling tankers up to 250,000 dwt, giving the port an export capacity of about 300,000 b/d.

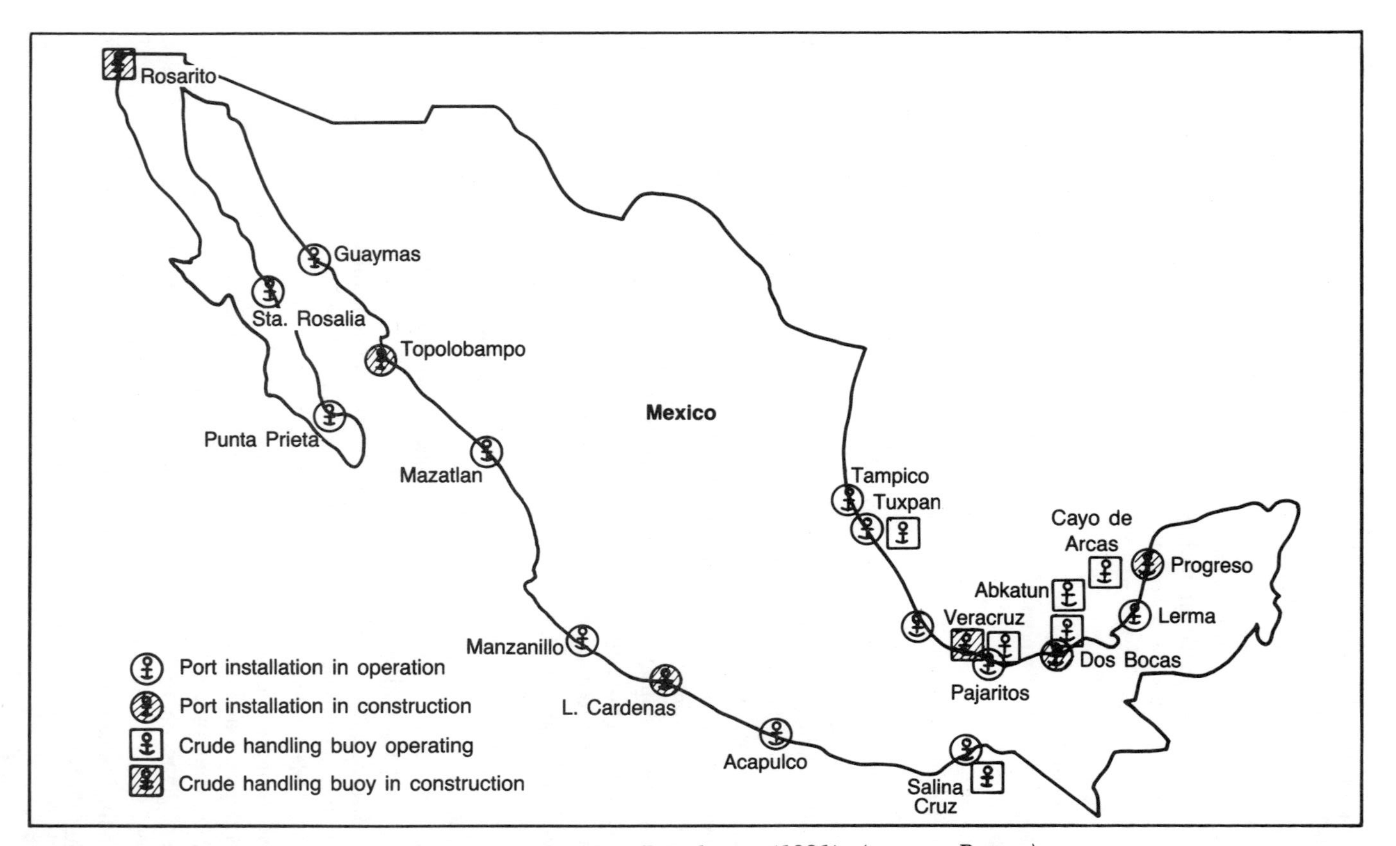

Fig. 3.3 *Where Pemex operates ports and crude handling buoys (1981), (source: Pemex)*

Pemex's plans for Dos Bocas, initiated in 1979, called for the port to have a storage capacity of 5.5 million bbl (11 tanks of 500,000 bbl each) and a pumping system of 60,000 bbl/hr to handle the crude originating from onshore fields. Pemex's plan called for the export of all onshore crude arriving at the port. Six tanks of 200,000 bbl each for the conditioning and storage of crude from Campeche Sound were also included in the plan. That gave Dos Bocas a total programmed storage capacity of 6.7 million bbl at the time of its scheduled completion in March 1981. Actual storage at the end of 1981 was 2.4 million bbl.

Salina Cruz, on the Pacific side of the Isthmus of Tehuantepec and site of a 170,000 b/d refinery completed in 1979, had been a transshipment point only for the domestic market during the 1976-81 period. By the end of 1981, most of the additions to port facilities had been completed for turning Salina Cruz into the main export terminal for the Orient and the U.S. West Coast. These improvements include crude storage and pumping capacity augmented to 2.5 million bbl, expansion of the oil pipeline from Nuevo Teapa to Salina Cruz, and installation of submarine pipelines and a buoy capable of loading and unloading tankers of 250,000 dwt.

In his speech at the inaugural ceremonies at Salina Cruz, held on 10 April 1981, President López Portillo noted that the new installations would bring about a 53% reduction in the distance travelled by crude exports from their production source in the Reforma and Campeche fields to markets in the Far East. By the end of 1981, Mexico had crude oil customers in Japan, Korea, and the Philippines.

The President added, with evident pride, that improvements and additions to port facilities at Salina Cruz had brought Mexico's installed export capacity to 3.75 million b/d.

Pipelines

Distribution of petroleum and its products within Mexico has long been a sticky problem for two reasons. First, as one commentator has noted, Mexico's terrain was created as God's museum of geographical extremities. Getting from one place to another is an engineer's nightmare and a mountaineer's dream. The Sierra Madre Oriental runs south from Texas for a thousand miles along the Gulf Coast, separating the Central Plateau from the coastal plains, and rises in the south to a monumental green wall 9,000 ft high *(see* Appendix I).

The second reason for the difficulty of the distribution of petroleum is that 96% of production takes place along the coastal plain or offshore in the Gulf, while in excess of 25% of consumption is in Mexico City and its environs and about another 25% is in the rest of the Central Plateau.

In 1938 domestic sales of oil products amounted to 17.2 million bbl, while in 1981 they reached 398 million bbl plus 1.372 bcfd of natural gas. From 1976 through 1981, the domestic sales volume of petroleum products increased by 141 million bbl, a 55% increase. Internal sales and exports of natural gas went to 1.675 bcfd in 1981, an 89% increase from 886.4 MMcfd in 1976. Such large and rapid increases required a major expansion in the country's transportation and storage capacity.

That expansion has been most noteworthy in the area of pipeline construction. In 1976, total pipeline length (gas, oil, and products) was 8,858 miles. By the end of 1981, that figure had reached 14,588 miles (4,719 miles products and petrochemicals; 6,277 miles gas; and 3,592 miles oil). The oil pipelines connect the major producing areas of Reforma-Campeche and Poza Rica with refineries along the Gulf Coast, at Salina Cruz on the Pacific and on the Central Plateau and at Monterrey.

The product pipelines link the refineries and the maritime terminals to the nine major regional storage and distribution centers spread throughout the country. Those centers and maritime terminals gave Mexico a storage capacity of 19.9 million bbl in 1982.

Natural gas is moved primarily through the National Gas System, which consists of three large interconnected gas pipelines. One line, Ciudad Pemex-Mexico-Guadalajara, connects the Reforma-Campeche area with the Central Plateau and is being extended to Manzanillo and Lazaro Cardenas on the Pacific. The second line, Cactus-Monterrey, connects the main gas production zone along the Gulf Coast to the industrial center of Monterrey. The northern region of Mexico is served by the third line, Reynosa-Monterrey-Torreon-Chihuahua, which added a new leg in 1981 to Ciudad Juarez (El Paso, Texas) to move Mexico's gas exports to the U.S.

The second line, also known as the Natural Gas Trunkline System, links the entire gas distribution system of the country. Started up in March 1979, it is 775 miles long and has a 48-in. diameter for 88% of its length. It took 17 months, 16 billion pesos, and 38 million man-hours to build. A private Monterrey company, Tamsa, manufactured 110 miles of the 48-in. pipe, with the remainder supplied by firms in Japan, Germany, Italy, and France. U.S. firms are glaringly absent from this list. The Mexico-U.S. dispute over natural gas during this period apparently did not help U.S. firms in their efforts to supply some of the pipe.

Pipelines are Pemex's most important mode of transport within Mexico, handling 75–80% of all crude oil, gas, and derivatives shipped during 1980. The remainder was divided equally between transport by ship and by the railroad and highway systems. The contribution of the highway system has grown at an annual rate of 23% during the 6-year period. Truck tankers increased their carrying capacity to 6.67 MMcf in 1981 from 2.36 MMcf in 1976. However, the capacity of the rail tankers has changed very little over the same period. Rail-tanker capacity dropped to 6 MMcf in 1981 from 6.8 MMcf in 1980, although the 1980 figure was up from 6.3 MMcf in 1976.

This chapter outlines only the investments of the government's portion of the petroleum sector.[3] Virtually no data exist for private-sector investments by either Mexican or foreign firms.[4] Statistics covering the industrial results of government investments, however, are abundant.

---

[3]In Appendix C Table C.1 gives Pemex capital spending figures for 1974–1981.

[4]In Appendix I Table I.2 gives a clue about private-sector activity in the retail distribution of Pemex-refined products.

4

Output of the petroleum sector

Crude Oil

Pemex's accomplishments during the first 5 years of the López Portillo period constitute a record of science fiction proportions. The crude oil production goal of 2.242 million b/d for 1982, established by the 6-year plan of 1976, was met one year early. Production rose to 2.746 million b/d in 1982 from 800,900 b/d in 1976, a 243% increase. In December 1982 Pemex reached the 3 million b/d mark.

The rise in crude production from the time the industry was nationalized in 1938 to 1981 progressed very gradually until the mid-1970s. The dramatic increase of the 1976–81 period reveals an average annual increase, in crude oil production, of 23.63% (Table 4.1). All of this increase came from the Southern Zone, while both the Northern and Central Zones fell back in production over the 6 years (Fig. 4.1). In 1981 the Southern Zone contributed 91.9% of total crude output, with the Central Zone adding 5.7% and the Northern Zone 2.4%.

The Campeche fields provided the major impetus to increased production in Mexico after 1976, accounting for 81% of all crude production increases during

Table 4.1
South dominates crude production

	(1,000 b/d)					
	1976	*1977*	*1978*	*1979*	*1980*	*1981*
Total	800.90	981.10	1,212.60	1,471.00	1,936.00	2,313.00
Northern Zone	59.50	67.40	69.90	63.10	55.00	54.40
Central Zone	137.20	126.80	130.80	124.10	128.20	133.40
Southern Zone	604.10	786.80	1,011.90	1,283.80	1,752.80	2,125.10
Tertiary	152.80	139.40	143.10	154.80	140.90	128.50
Cretaceous	451.30	647.40	868.80	1,077.00	998.50	914.10
Campeche	—	—	—	51.70	613.40	1,082.50

Source: Pemex

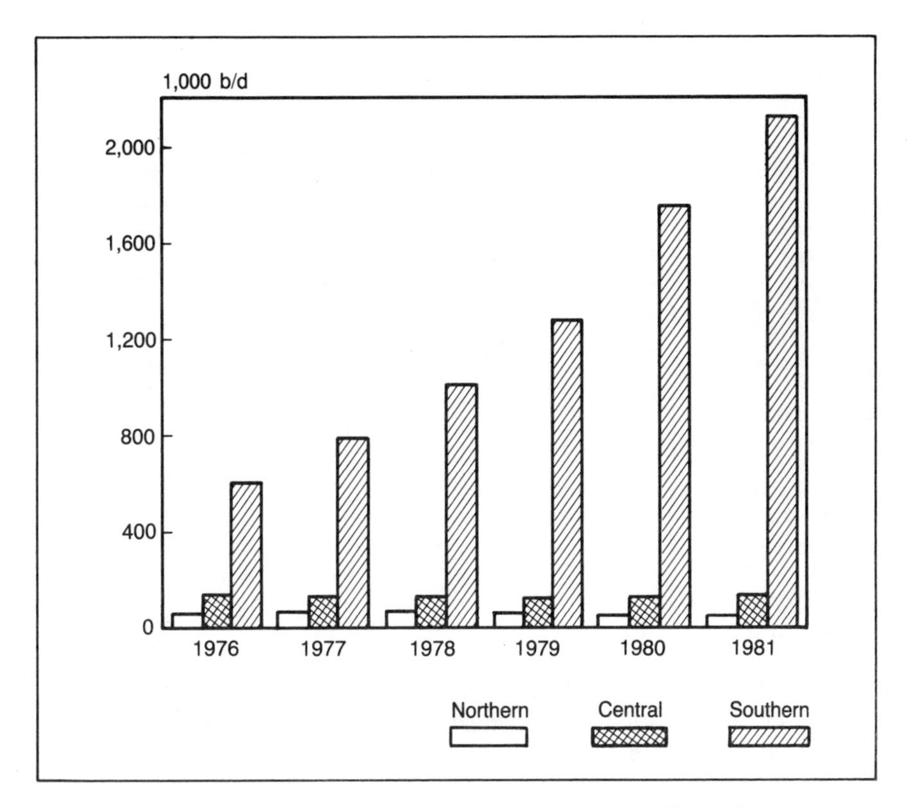

Fig. 4.1 *Southern zone leads crude production (source: Pemex)*

the period. The other significant area of growing output was the Mesozoic Chiapas-Tabasco (Reforma), which pumped 39.5% of the crude produced in 1981. A critical contribution from this area was made by the Huimanguillo fields, which came on stream in 1977–78 and yielded over 200,000 b/d of super-light crude by 1981.[1]

Natural Gas

Production of natural gas in Mexico experienced a steady increase from the time of nationalization of the industry through 1981. As in the case of crude oil production, the 1976–81 period showed a precipitous growth curve. Total production nearly doubled during the period, jumping to 4.0608 bcfd in 1981 from 2.1086 bcfd in 1976. The original Pemex 6-year plan called for gas production to reach 4 bcfd in 1982; so as with crude production targets, Pemex came in one year ahead of schedule (Table 4.2).

[1]See also Appendices F and M.

Table 4.2
How Mexico's gas output climbed

			(MMcfd)			
	1976	*1977*	*1978*	*1979*	*1980*	*1981*
Total	2,109	2,046	2,561	2,917	3,548	4,061
Associated gas	1,059	1,190	1,624	2,210	2,578	3,046
Northern Zone	77	83	109	104	81	86
Central Zone	151	133	159	183	187	168
Southern Zone	831	975	1,355	1,923	2,311	2,792
Tertiary	198	154	150	153	137	130
Cretaceous	633	821	1,205	1,747	1,915	2,133
Campeche	0	0	0	23	258	530
Nonassociated gas	1,050	856	938	706	970	1,014
Northeast District	383	391	523	560	519	456
Reynosa	—	—	413	391	369	356
Monclova	—	—	110	169	150	99
Poza Rica district	0	0	0	0	0	17
Papaloapan	23	23	22	22	22	32
El Plan district	0	0	0	0	0	5
Ciudad Pemex	644	442	392	124	429	506

Source: Pemex

The most striking change during the 6-year period, other than the production growth factor, was the improved utilization of natural gas. In 1976, nearly 500 MMcfd was flared, or almost one-fourth of total production. Through the installation of 693,620 hp of gas compression between 1977 and 1981, only 16% of total gas production was being flared by 1981. Discounting the Campeche fields, Pemex asserts that 98% of the gas produced onshore was utilized in 1981.

The driving force behind higher levels of gas production was higher levels of crude production. Production of nonassociated gas was lower in 1981 than in 1976, although it did show steady growth after bottoming out in 1979. Therefore, all of the overall increase in production of natural gas came with the associated gas produced with crude oil. Many of the fields discovered during 1976–81 in the Mesozoic Chiapas-Tabasco area were producers of volatile light oil with high contents of natural gas.

Associated gas in the Southern Zone accounted for all of the growth during the 1976–81 period, with 87% of all associated gas production in the period and 92% in 1981. In 1979, Campeche began to contribute to gas production, although until November 1981 all of its gas was flared.

Petroleum Products

In 1938 Pemex had six refineries with a capacity of 114,000 b/d. Mexico's refining capabilities in the 1938–76 period underwent a very gradual expansion.

The average annual growth rate for each decade from 1938–68 stayed between 5% and 6% and then dropped to 4.5% during the 1968–75 period. From early 1976 to late 1981, refinery capacity went to 1,523,500 b/d from 785,000 b/d, a 94% increase. Covering the same period, total runs went from 739,700 b/d to 1,271,900 b/d, a 72% rise. By October 1980 Pemex had become the fifth largest refinery company in the world.[2]

In 1981, 600,000 b/d of capacity was under construction, and plans to start construction of a new refinery in 1982 were delayed indefinitely because of budgetary restraints imposed in the wake of the peso's devaluation in August 1982. Additions to capacity underway at existing installations in 1981 were divided as follows: Madero, 150,000 b/d; Salina Cruz, 300,000 b/d; and Tula, 150,00 b/d.

The Central Plateau, including Mexico City, is supplied by the refineries at Azcapotzalco, Tula, and Salamanca. On the eastern coastal plains the refineries of Ciudad Madero, Minatitlan, Poza Rica, and Cactus meet the total demand on the Gulf Coast and support supplies to the Central Plateau and the Pacific Coast. The recently built refineries at Tula (1976), Cadereyta (1979), and Salina Cruz (1979) are part of a government effort to decentralize industry and employment. These last two refineries, both completed one year behind schedule, were built with Mexican technology, basic engineering, and licensed processes.

Heavy crude reserves in the Sound of Campeche and increased production of those highly productive fields have made it necessary to process greater volumes of heavy crude in Pemex refineries. In 1981, Pemex planned to process runs of 40% Maya (heavy) crude and 60% of other crudes as an average for the year—that is, 490,000 b/d of Maya. Pemex managed to reach an average of 231,000 b/d of Maya, 20.6% more than in 1980 but far short of its target.

As a result of this continuing shift toward heavier crude runs, each average barrel processed will yield more residual fuels and other heavy products and less gasoline and light products. This marketing difficulty is the dark cloud around the silver lining of the large heavy crude reserves and prolific production coming from Campeche fields. The end was nigh for the 1976–81 period when the problem of higher participation of heavy crudes in refinery runs began to hurt.

Resolution of the problem rests upon refining and marketing policies implemented in the post-1981 period. As an interim measure, Pemex changed its strategy in the second half of 1981 to refine a greater percentage of light crudes, forcing more heavy crude than planned into the export market.

Production of gasoline and diesel since 1938 has climbed without pause, while that of residual fuel has staggered a bit on the way up (Fig. 4.2). The expansion of Mexico's refining capacity and production has come in response to the steady growth of internal demand for refined products, which averaged 7%

---

[2]See also Appendix G.

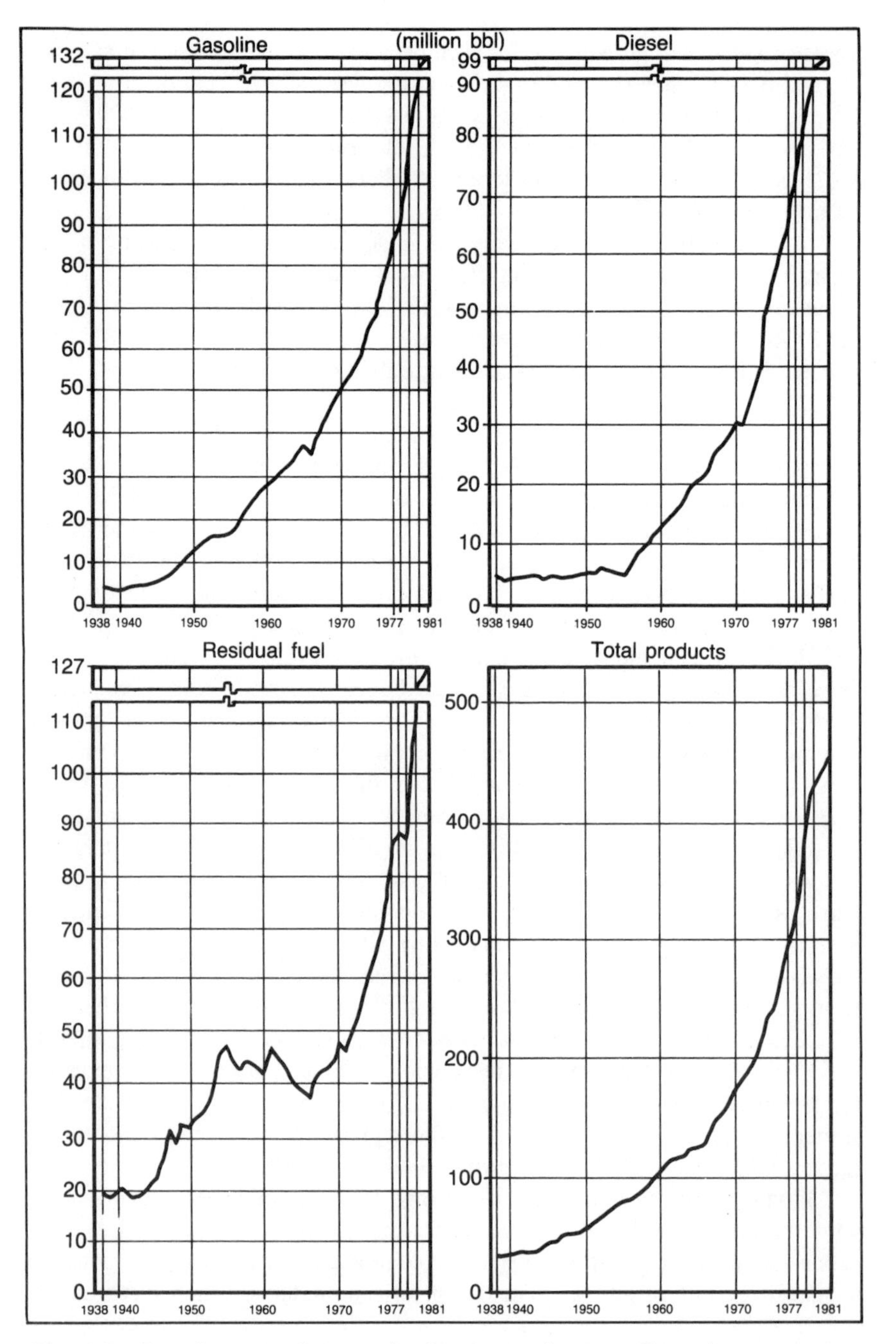

Fig. 4.2 *Petroleum product production jumps (source: Pemex)*

during 1976–1981. Particularly large increases from 1976–1981 were reached in the production of liquefied gas (143%) and gasoline (70%).

Output of Petrochemicals

Processing of natural gas

The growing production of crude oil during 1976–81 also meant sharp increases in gas processing to handle the bigger volumes of associated natural gas (Table 4.3).

Table 4.3
Crude oil and associated natural gas production

	1976	*1977*	*1978*	*1979*	*1980*	*1981*	*Average Yearly Growth*	*Percent Variance 1976–1981*
Crude oil (1,00 b/d)	800.9	981.1	1,212.6	1,471.0	1,936.0	2,313.0	23.6	188.8
Associated natural gas (MMcfd)	1,059.1	1,190.4	1,623.9	2,210.2	2,578.0	3,046.4	23.5	187.6
Gas-oil ratio	1.32	1.21	1.34	1.50	1.33	1.32		

Source: Tables 4.1 and 4.2

That growth led to the installation of new gas processing units at the petrochemical complexes at Cactus in Chiapas and Poza Rica in Veracruz.

In fact, Mexico in 1982 had among the largest-capacity installations in the world for the recovery of natural gas liquids. The plant at Cactus had a capacity of 1.6 bcfd, while La Venta and Ciudad Pemex in Tabasco had capacities of 387 MMcfd and 733 MMcfd, respectively.

Gas processed by Pemex has jumped to 3.241 bcfd in 1981 from 1.395 bcfd in 1976—a 232% increase. The recovery of gas liquids during the same period grew by 157%, to 239,777 b/d from 99,355 b/d.

These advances were made possible by the addition of 21 new plants (7 cryogenic, 11 sweeteners, 1 stabilizer/sweetener of condensates, and 2 gasoline fractionators). By the end of 1981, Pemex operated 16 sweetener plants with a processing capacity of 2.960 bcfd and 15 cryogenic absorption plants, for a total processing capacity of 3.957 bcfd with an average recovery of 242,000 b/d of gas liquids.

Total average annual growth for the period was 13.57% with three of the six plants—Cactus, Ciudad Pemex, and Poza Rica—accounting for all of the increase. Capacity at the end of 1981 was only 9% more than net production of natural gas plus the quantity of gas flared.

Basic petrochemicals, 1976–1981

Since 1958, Mexico's petrochemical industry has been divided into two main areas: basic and secondary petrochemicals. Government regulation of this industry is based upon "Segundo Articulo de La Ley Reglamentaria del Articulo 27 Constitucional en el Ramo del Petroleo," published in the *Diario Oficial*.[3] It reserves to the nation, through Pemex, the manufacture of basic petrochemical products including olefins, aromatics, ammonia, and any other potential feedstock. Article III of La Ley Reglamentaria establishes the government as the authority for determining which products fall under the classification of "basic" petrochemicals.

The basic petrochemical industry in Mexico began in 1951 with the production of sulfur in the Poza Rica plant at Veracruz. Real growth in the industry did not take place until the 1960s with the opening of ammonia plants at Salamanca and Cosoleacaque and the aromatic system at Minatitlan, Veracruz.

A jump in the production of basic petrochemicals occurred between 1976–1981. Production in 1981 was bumping up against the 10.14 million short-ton level, compared to 4.3 million tons in 1976—or 2.5 times greater in 1981 than in 1976. In 1982, production reached 11.22 MM tons, which came very close to the López Portillo program goal of tripling output during the sexennium. This rapid growth soon led to petrochemicals being called a "wonder" industry in Mexico, though some wit commented that it was because "one wonders how to pronounce the name of the chemicals and then wonders what they are used for."

The overall annual growth rate for basic petrochemical products from 1976 to 1981 was 18.34%. Methanol was the growth leader with an annual growth rate of 41.04%, with ammonia, acrylonitrile, and acetyldehyde all about 20%/year. Such rapid growth was accomplished with the addition in this period of 25 petrochemical plants and 21 auxiliary processing units. On the average during the 1976–81 period, a new petrochemical plant was started every 45 days.

In Pemex's own evaluation, its 5-year program (1977–1981) achieved 93% of its goals in the petrochemical sector, the shortfall resulting from the delay of some programmed works originally slated to start up by 1980 but deferred for budgetary reasons. The operation of a group of plants that make up a part of La Cangreja Complex with a planned yearly production volume of 3.5 million tons was set back by the financial crisis.

As 1981 ended, Pemex counted on a nominal capacity of 12,897,027 tons/year and supplied a range of 41 basic petrochemical products, which satisfied approximately 85% of national demand.

Pemex also added new products to its line of basic petrochemicals in the past five years, including ethylene perchloride, high-density polyethylene, oxygen, and cumene.

---

[3]*Diario Oficial*, 9 February 1971.

One of the most salient events of 1981 in the industry was the commencement of operation of some of the new plants at La Cangrejera Complex. Production began of acetic acid, ethylene oxide, and cumene, which should eliminate exports of these products.

In Cosoleacaque, Veracruz, May and October saw the initiation of operations at ammonia plants 6 and 7, each with capacities of 490,528 tons/year, making it the world's largest center for the production of ammonia.

Mexico is not resting on its petrochemical laurels. During the first 6 months of 1982, Pemex's 40 basic petrochemical plants produced 5.4 million tons, a 20% increase from the first half of 1981. Early in 1982 before the onset of the financial crisis, Pemex had 20 new petrochemical plants under construction, each with a scheduled start-up date of December 1983. These would have had an installed capacity of 3.3 MM tons/yr, increasing Pemex's overall capacity to 16.1 MM tons/yr. Work on these plants plus 40 others slated for construction in 1983 was virtually suspended as a result of the foreign-exchange crisis.

Nevertheless, Mexico is still one of the world's petrochemical growth leaders. The de la Madrid government is expected to promote an import substitution program aimed at increasing domestic production of petrochemicals currently being imported.[4]

The story of how Mexico got oil and gas out of the ground and into storage tanks and pipelines is a tale of costs. Sales, domestic and export, required additional direct and indirect costs. However, the story of sales is the other shinier side of the coin.

---

[4]See also Appendix H.

5

Sales of Pemex

Domestic Sales Volume

Unfortunately, the total sales of the petroleum sector, including the sales of oil-field equipment and related services in Mexico, has never been tabulated. Nor has the market value of the services performed by Pemex or its subsidiaries abroad, in Cuba for example, ever been published. What can be discussed is product sales by Pemex. The volume of petroleum product sales in Mexico, all of which are handled through Pemex and its distribution agencies, grew at an 8.5% annual rate from 1976 through 1981 (Table 5.1). This pace slightly exceeded the economy's real GDP growth rate of 7.4% for the same period.

The sales growth of four principal products was particularly sharp during the decade of the 1970s when the Mexican economy entered its period of rapid

Table 5.1
Pemex product sales expansion

			(1,000 b/d)				*Ave annual*
	1976	*1977*	*1978*	*1979*	*1980*	*1981*	*growth, %*
Gasoline	208	219	238	276	314	358	11.51
LP gas	63	62	77	87	103	114	12.46
Kerosene	51	53	55	61	65	67	5.41
Diesel	164	169	188	203	215	233	7.32
Residual fuel	195	196	229	223	243	250	5.05
Natural gas*	135	137	161	198	203	209	9.17
Total	815	836	948	1,048	1,143	1,230	8.57

*Equivalent to residual fuel
Source: Pemex

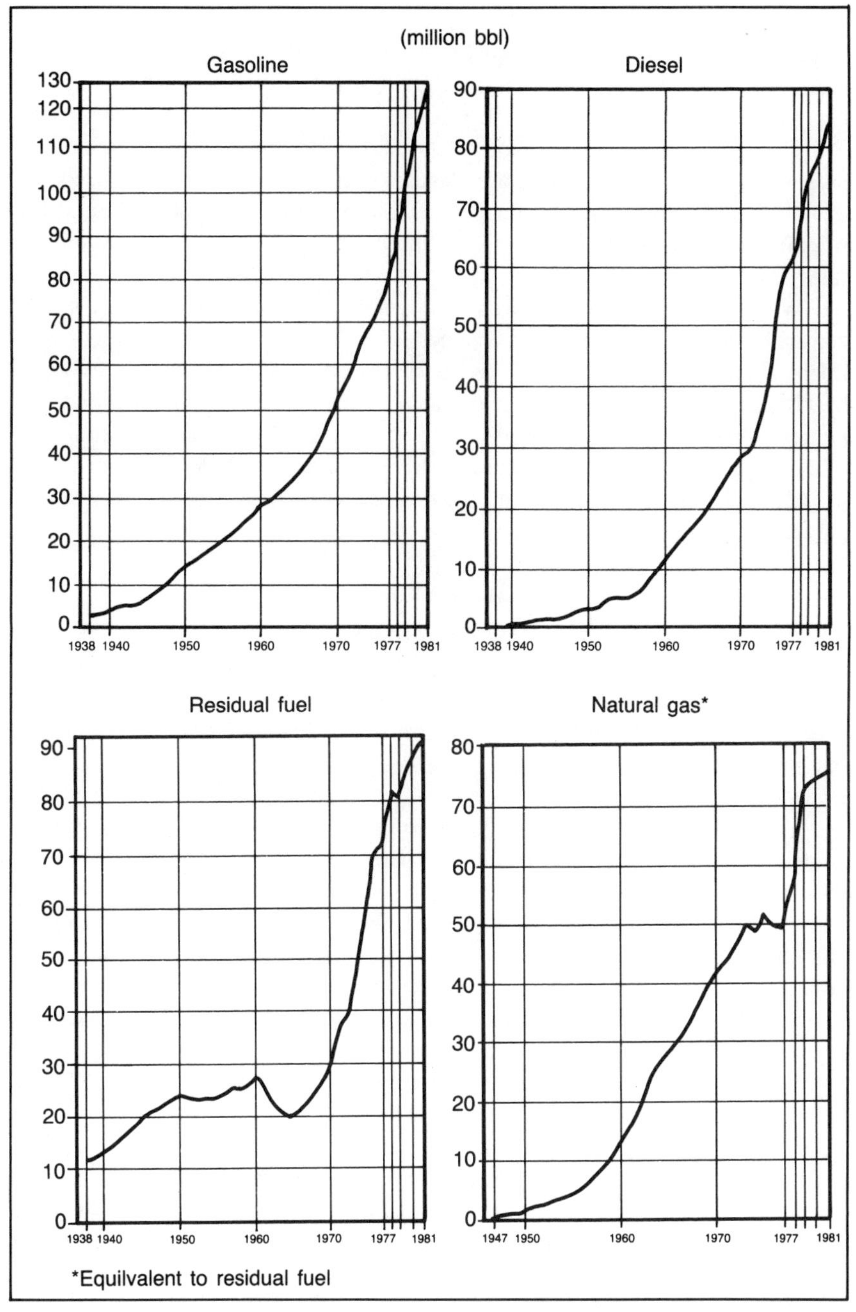

Fig. 5.1 *Domestic product sales soar (source: Pemex)*

growth (Fig. 5.1). Most noteworthy was the growth of natural gas sales, which received a strong and abrupt impetus to sales in 1977 from increased crude production and the associated natural gas produced as a result. Sales of natural gas grew at 9.1% through the 1976–81 period.

This rapid growth rate demonstrates the success met by the government's program to encourage use of natural gas by industry and the energy sector, thereby freeing more crude oil for export. Industrial consumption of natural gas went from 640.2 MMcfd in 1976 to 1,026.7 bcfd in 1981, a 9.9% annual growth rate.

What does not show in the sales figures is the petroleum sector's own consumption of natural gas, for which Pemex does not receive any income or financial credit. Pemex's natural gas use increased 142% from 1976 to 1981 and by 1981 consumed more than 42% of total available natural gas production. This Pemex consumption in 1981 had a value on the U.S. market for natural gas, based on prices to industrial consumers, of $1.487 billion (452.125 bcf at $3.29/MMcf). Sales figures alone, then, drastically underestimate the productive role played by natural gas in Mexico's economic activity.[1]

Sales Income

Pemex reported early in 1983 that domestic sales from 1977 to 1982 amounted to 520 billion pesos, of which 81.2% was represented by petroleum products, 16.3% by petrochemicals, 2.1% by natural gas, and the remainder by other products. Of greatest interest here is not how much Pemex earned on sales, but how much it and, in turn, the national treasury lost due to subsidized price levels.

The 1981 sales of three important products have been used to illustrate the value and magnitude of government price subsidies (Table 5.2). The volume and the value of sales of each product are those provided by Pemex, with values converted from pesos to dollars at the average official exchange rate for 1981. Although other measures could be used to estimate the subsidy value, this comparison is made against U.S. product prices in December 1981. One can argue that, as in the case of government-regulated natural gas prices, the U.S. price is not the world market value. The method simply compares the income generated by sales of those products in Mexico with the income that would have been generated by their sale in the more open U.S. market.

For gasoline this meant that Mexican prices would have had to increase by a factor of 5 to equal its U.S. value, and for natural gas and residual fuel by factors of 6.2 and 8.2, respectively (Fig. 5.2). For just 1981, total subsidies for these three products (by this measure) amounted to $9.27 billion, representing 64% of Pemex's income from exports.

[1]See also Appendix J.

Table 5.2
How Mexico subsidizes domestic products (1981)

	*Gasoline** *(Billion gal)*	*Natural gas†* *(Bcf)*	*Residual fuel‡* *(1,000 bbl)*
Volume	5.4	517	91,104
Reported value§ (Billion 1981$)	1.4	0.27	0.31
U.S. market value (Billion 1981$)	7.0	1.7	2.5
Implicit subsidy (Billion 1981$)	5.6	1.4	2.2

*December 1981 average U.S. price for leaded regular, $1.29/gal; unleaded premium, $1.43/gal (based on 96% of Mexican sales of regular and 4% premium)
†Dec. 1981 average U.S. price for natural gas to industrial customers of $3.29 Mcf
‡Residual fuel #6 U.S. price of $27.69/bbl
§At average official exchange rate of 24.51:1 U.S.$
Source: Prices from *Energy Detente,* 21 December 1981; Pemex

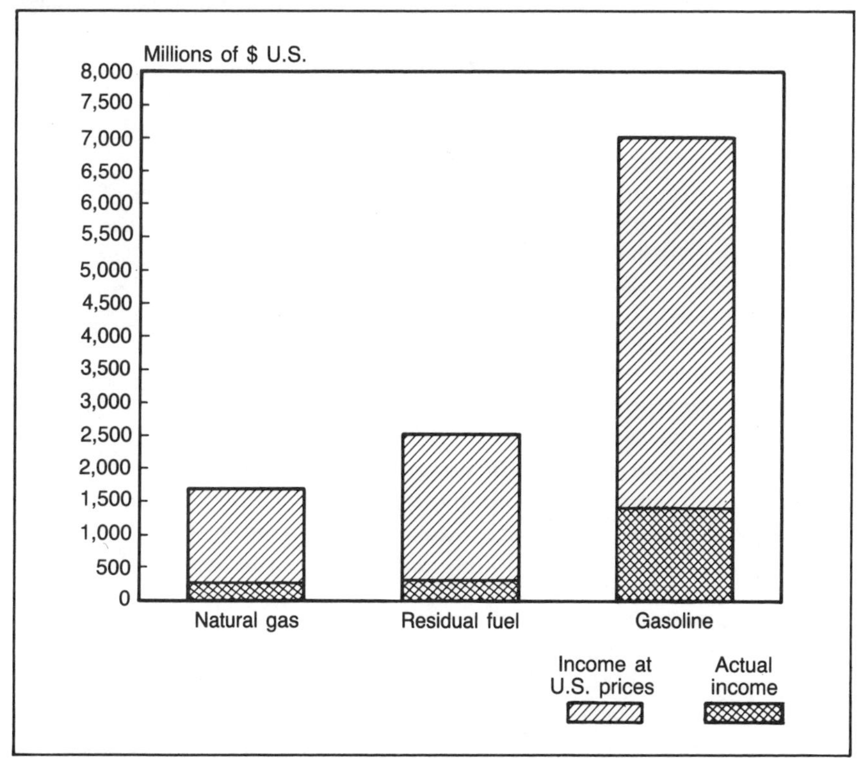

Fig. 5.2 *Pemex loses potential income (selected products: subsidy 1981), (source: Baker & Associates)*

Crude Export Sales Negotiations

Throughout the oil-scarcity years of the López Portillo period, Pemex maintained crude export prices equal to or higher than OPEC prices. Mexico reasoned that security of supply, among other factors, was worth a premium price. Obtaining or increasing the volume of a crude contract with Pemex has been for many companies—including some of the majors—a hand-wringing experience. Any number of companies sent their presidents and chairmen of the boards to the office of the Director General of Pemex, hoping in this way to show the importance of a Pemex contract for the company in question.

As a rule, Pemex sold to its regular customers, and woe to anyone who may have had a contract in the past who either did not lift or who cancelled the contract entirely. "Mexico has a long memory" was what such fair-weather buyers were told, and, in saying this, Pemex showed its basic attitude toward export sales: the customer was really not buying oil, in the American sense of the term, but was being given the privilege or concession of acquiring Mexican oil through the vehicle of a commercial contract. All such concession-holders were told that they must not abuse their concession by selling their oil to third parties—and, by definition, no oil whatsoever was to be sold to trading concerns. Mexicans felt that they needed to know the actual industrial need of the customer before they could allow him to acquire rights to part of the national patrimony (here measured in barrels of oil per day).

Where the credentials of the buyer were suspect—and companies often went to great length to document the need of their refineries for Mexican oil—or where ulterior, trading motives were thought to be present, buyers were turned down. A letter from the Commercial Subdirector of Pemex (read: Vice-President, Sales) might read:

> *In accordance with our conversation . . . in our offices, I have asked the Foreign Commerce Group to review your requirements in purchasing crude oil for your processing operations. Unfortunately, all the volume we have available for export is placed in this moment under long-term contracts; consequently, it is impossible for Pemex to offer you any volume. If by any reason the above-mentioned situation changes, we will be most willing to talk again with you.*

It was commonly believed that Pemex itself had little authority in the matter of oil "allocations," that all contracts exceeding 10,000 b/d needed the approval of the President of Mexico himself. The letter just quoted refers to a Foreign Commerce Group, which was an advisory body that made recommendations on oil export policy in general and in particular as it might affect a specific company. Such advisory bodies were always in existence, although their names and membership might change.

Under López Portillo the Energy Commission (with its own technical secretariat) was a body that included several cabinet members. Under de la Madrid this mechanism continues but under a different name. *Platt's Oilgram News* of 28 December 1982 reported the establishment of a Foreign Trade Committee, headed by Pemex president Mario Ramon Beteta. The purpose of the committee was to "coordinate and monitor aspects of crude, products and petrochemicals marketing." Committee members, at the time the committee was established, included the Undersecretary for Economic Affairs in the Foreign Relations Ministry, the Undersecretary of the Budget and Programming Ministry, the Undersecretary of Foreign Trade in the Commerce and Industrial Development Ministry, and Pemex's own Commercial Subdirector.

Another move of the de la Madrid team was to appoint a new officer in the Pemex Sales Department with the title of International Trade Coordinator. (The person appointed previously had been the head of the technical secretariat of the Energy Commission.) Through the mechanism of such labyrinthine bureaucracies, Mexico achieves here, as elsewhere, effective obfuscation of decision-making authority.

In response to this confusing bureaucracy, a new breed of consultants emerged—mainly in Mexico, but also in the U.S.—who promised "contacts" in the upper reaches of Mexico's decision-making ivory tower. These consultants offered various credentials, ranging from a portfolio of photographs showing the consultant playing tennis with the President to the claim that someone was and had been on a first-name basis with so-and-so for years. As is generally the case in Mexico, many people know or are related to the proverbial "cousin of the President," with the result that some companies got themselves involved with bizarre connections that typically led nowhere.

As for the professional staff in the export office of Pemex's sales department, there were never even rumors that a company had been able to gain an improper personal influence over crude export contracts. In this area, the ship was run very tight because the persons in these offices were contract administrators and not contract decision-makers in their own right.

At the other extreme, there was at least one instance during the López Portillo period in which a U.S. oil company senior executive tried repeatedly to have oil allocations settled in his company's favor through direct contacts with the President. The executive, owing to his own very high contacts, was indeed able to meet with the President on a number of occasions, but in this particular case the personal touch seemed to backfire.

Mexican nationalists feared that if the outcome of such presidential politics were to result in oil contracts, two unfortunate results might occur. First, the name of the President might be clouded in the public's mind as someone who, for personal gain or for having been intimidated by foreign interests, had authorized an oil contract in a particular case. For this reason, the locus of

Table 5.3
Mexico diversifies crude exports

	(1,000 b/d)						
	1976	*1977*	*1978*	*1979*	*1980*	*1981*	*1982/Oct*
U.S.*	73.50	178.50	324.90	448.80	562.50	546.70	709.17
Israel	20.70	20.20	22.00	40.80	56.60	64.80	71.54
Spain		2.40	13.60	42.90	92.50	151.50	161.36
Canada		0.90	2.40	—	4.20	46.10	48.04
Japan			0.90	—	35.20	76.50	101.14
Holland			1.20	0.30	—	—	—
Costa Rica†					4.90	5.40	4.28
France					42.10	71.70	80.25
Yugoslavia					3.10	0.90	—
Nicaragua†					2.30	5.50	7.69
El Salvador†					0.50	5.70	4.97
Brazil					16.80	51.70	55.63
Bermuda					7.00	1.00	—
U.K.						18.30	78.36
Dominican Republic†						9.60	12.13
Panama†						8.20	13.22
Jamaica†						7.50	6.84
Philippines						7.20	8.57
Guatemala†						5.80	4.80
Korea						4.80	15.39
India						2.50	—
Switzerland						2.30	—
Haiti						0.80	—
Italy						1.00	33.89
Honduras†						0.50	—
Colombia						0.40	—
Portugal						1.60	12.17
Uruguay							8.35
Rumania							3.44
Austria							6.02
TOTAL	94.20	202.00	365.00	532.80	827.70	1,098.00	

*Includes Puerto Rico
†Pact of San José countries. Total shipments (M b/d) were 7.7 in 1980 and 49.9 in 1981.
Source: Pemex

authority in oil contract matters, which in many if not most cases led from Pemex's offices on Marina Nacional to the President's headquarters in Los Pinos, was never made explicit or visible.

The second reason why Mexicans did not want to see the outcome of executive-to-President talks to result in oil contracts was because of what they feared might happen should the executive speak indiscreetly to the U.S. press, conveying the impression that this executive might be authorized to act as a private-sector spokesman for Mexico in the U.S. In short, Mexico wanted to avoid implicating the President in oil deals of any kind to avoid giving any one person or company the presidential blessing.

Pemex had developed a diverse portfolio of company and country buyers in the López Portillo era (Table 5.3). The big company buyers were Exxon and Shell. Some of the companies on the list dropped out prior to or in the course of the 1981 oil-price fiasco in which Pemex went from June's export level of 1.1 million b/d to 457,000 b/d in July (Fig. 5.3).

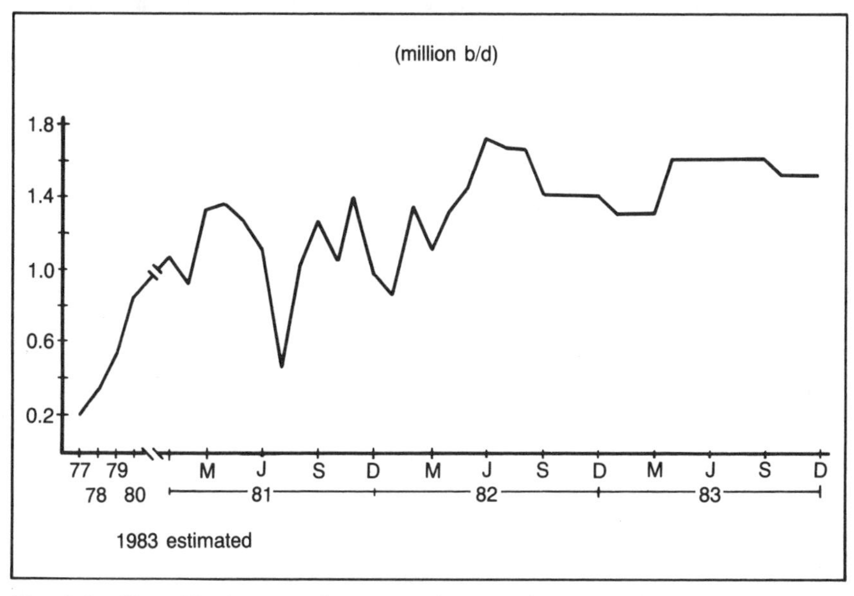

Fig. 5.3 *How Mexican crude exports increased (source: Pemex)*

A central matter affecting export sales was the famous "mix" that Mexico imposed on its crude customers. Mexico required that customers lift 50% heavy Maya oil as part of any lifting contract for light Isthmus oil. Many refineries were simply not equipped to use the heavy, high-sulfur Maya oil and either chose not to lift from Mexico at all or chose to exchange it, very quietly, for oil or dollars in the hands of other companies.

During 1979, 1980, and the first half of 1981, the López Portillo government became increasingly intractable about this requirement. In the months preceding the July fiasco, Maya exports jumped to much higher levels than Isthmus exports, while customers wanted just the opposite situation (Fig. 5.4). The crash of Mexican crude exports in July of 1981 was therefore not only a matter of price. Companies had to agree to lift both grades of oil, preferably at the same time. Later, more flexibility was observed and companies were allowed to lift Isthmus one month, Maya the next. At the very end of the period when Maya, for companies that could process it, was the best buy in the international oil market at $25/bbl, there was talk in official circles about the separate marketing of the two oils. Such talk would have been heresy in 1977–80.

Another factor affecting export sales was infrastructure, which, during the López Portillo period, was built up to handle about 2 million b/d. At the beginning of the period all exports were out of the Gulf, but in 1982 the crude pipeline to the Pacific Coast was completed, allowing Mexico to ship directly to its Pacific customers, mainly Japan.

A third element affecting exports was weather. The average month-to-month variation during 1975–80 shows exports dropping in winter months. Exports were up in the fourth quarter but down in the first, reflecting weather conditions in the Gulf area (Fig. 5.5). The beefing up of export port facilities in 1982 will make exports less vulnerable to seasonal weather conditions.

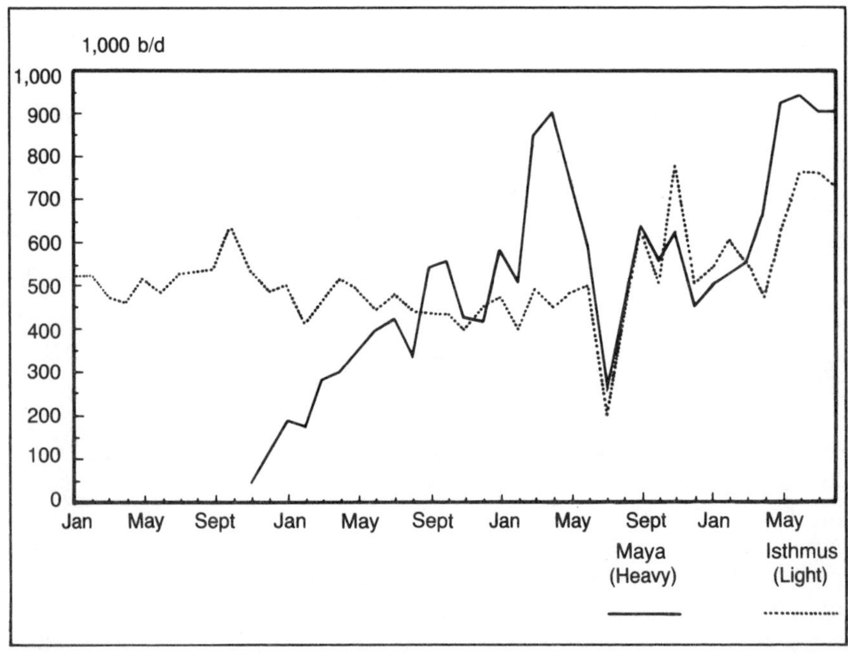

Fig. 5.4 *Maya gains prominence in crude exports January 1979–August 1982 (source: Pemex)*

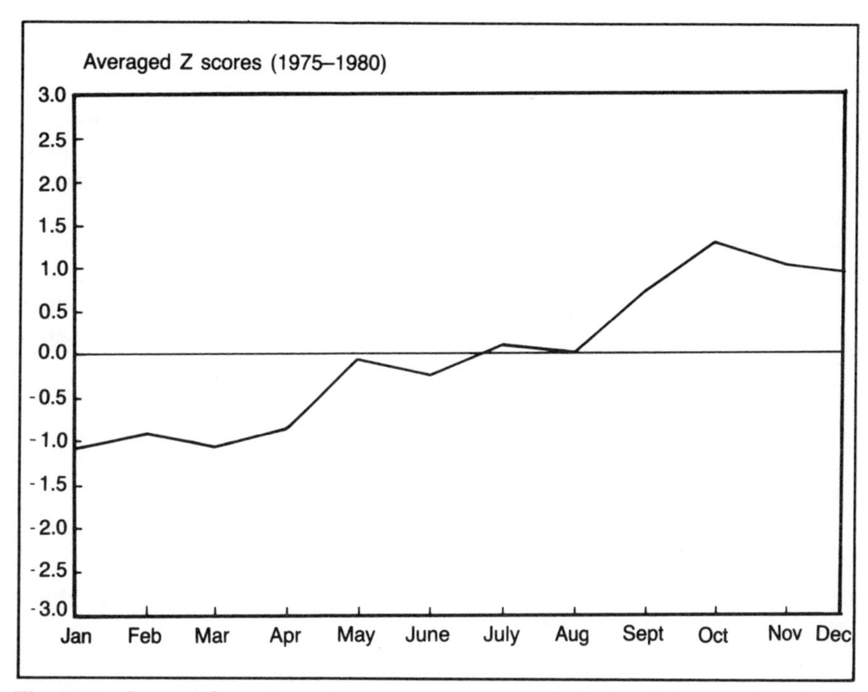

Fig. 5.5 *Seasonality of crude exports (z scores calculated from 1,000 b/d), (source: Baker & Associates)*

The López Portillo period also saw a drive to diversify its crude oil client portfolio. The reasoning of the Energy Program of November 1980 was that no country should become dependent on Mexican oil and, by implication, Mexico should not become dependent on any one buyer. Diversification also had the allure of "leverage," meaning that Mexico might be able to negotiate a package (perhaps the most abused word of the 1970s) involving foreign investment and technology in the context of oil sales.

This policy did have teeth in at least one instance. During the free-for-all in the summer of 1981, France announced that it would not lift Mexican oil during the third quarter. Suddenly, France was informed that, given the implicit understanding that the oil contract was granted in the context of—not in exchange for—French investments, technology, and complementary contracts, the cancellation of the oil contract carried with it the necessity to consider French bids and contract work (valued at a billion dollars) withdrawn or suspended.

Mexico boasted that political coloration never influenced oil contracts, in response to which skeptics pointed to Cuba, which by logistical sense should be receiving oil from Mexico instead of the Soviet Union. The explanation given was that Mexico could never get the Russians to agree on commercial terms, but die-hards remained convinced the reason was either U.S. pressure or fear of U.S. retaliation.

A quite different instance was South Korea, which was never the darling of Third World countries given its poor reputation for political liberty under President Park. It was only after very patient efforts, facilitated probably by the unexpected death of Park, that the deal was finalized. The closed-door history of export diversification under López Portillo is hardly revealed by a mere list of country clients (Table 5.3).

With the oil glut of 1981—caused by Saudi Arabia's sustained production of 9–10 million b/d coupled with the catching-up effect of fuel conservation measures in industrial countries—Mexico's best-laid plans, like those of all other oil-keyed strategists, were dashed. The lone survivor of Mexico's Energy Program of 1980 was the policy of limiting crude exports to 1.5 million b/d. With the oil glut, Mexico could defend its right to at least this export level in the face of OPEC brow-furrowing.

A paradoxical result occurred in May 1982: Mexico, despite its efforts to "diversify" its exports, surpassed Saudi Arabia as the principal foreign crude supplier to the U.S. Mexico achieved its objective of reducing the U.S.'s share of crude exports from the 80% range in 1976 to less than 50% in 1982 (Fig. 5.6).

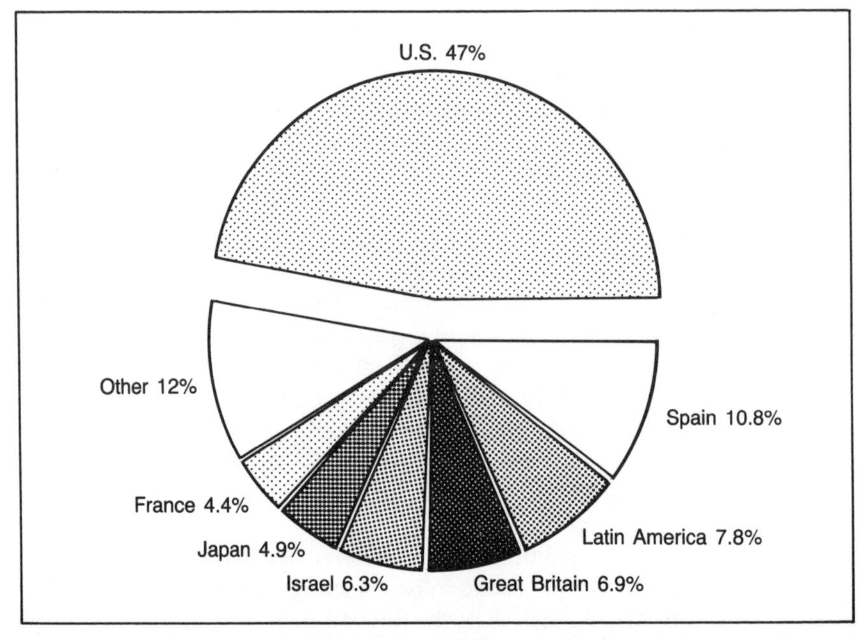

Fig. 5.6 *Crude exports by destination (1982 second quarter), (source: Pemex)*

Other Petroleum Trade

Production of refined products between 1976 and 1981 kept pace with internal consumption, with a very few minor exceptions in 1976 and 1978. Although trade statistics show Mexico importing several refined products, the value of direct exports exceeds the value of product imports.

Between 1976 and 1981 exports of basic petrochemicals showed a sharp increase in absolute terms and generally rose as a percentage of imports of petrochemicals (Table 5.4).

Table 5.4
Basic petrochemicals balance in current account

	(Millions of current dollars)					
	1976	*1977*	*1978*	*1979*	*1980*	*1981*
Exports	0.3	3.3	67.6	107.7	125.3	153.6
Imports	103.8	156.5	163.5	331.6	522.9	523.5
Exports/Imports %	0.3	2.2	41.4	32.5	24.0	29.3

Source: Pemex, *Memoria de labores* and Jaime Corredor, *"El petroleo en Mexíco,"* 1980.

In 1981, Mexico requested tariff-exempt treatment in the U.S. under the Generalized System of Preferences (GSP) for six petrochemical products (Table 5.5). Of these products, the first two are big-volume items in the U.S. market. Acetone is a byproduct of many chemical processes and acetonitrile is an important raw material in a number of secondary petrochemicals.

Table 5.5
Mexican requests in 1981
for GSP treatment of petrochemicals

Product	*Column 1 Tariff*	*TSUS No.*[a]
Acetone	1.01c/lb + 18.7% ad. valorem	404.64
Acetonitril	0.09c/lb + 20.5% ad. valorem	405.60
Benzyl alcohol	13.3% ad. valorem	403.45
Benzyl chloride	0.05c/lb + 12.5% ad. valorem	402.56
Benzyl dichloride	17.7% ad. valorem	402.80
Heterocyclic compounds	8.4% ad. valorem	406.36

[a] Trade Schedule of the U.S.

GSP authorities, with the concurrence of the semi-independent International Trade Commission, ruled that Mexico enjoyed built-in export competitiveness because of subsidized petrochemical feedstock prices. The price factor combined with falling demand and production in the depressed U.S. market provided sufficient reasons for the U.S. to refuse Mexico's request for GSP treatment on all six petrochemical products.

Future Export Levels

By the end of the López Portillo administration, the extent to which the world oil market had turned from one of scarcity to one of surplus had become an urgent problem for Mexico as well as for other oil exporters. The Energy Commission of the Institute of Economic and Political Studies of the PRI (Iepes), headed by the subsequently named Minister of Planning and Programming, prepared a briefing paper for the president-elect regarding the incentives and constraints facing Mexico in the years ahead. The report, called "Energy and the External Sector," received wide attention in the U.S. press. Such "leaking" of government documents to the foreign press is quite uncharacteristic of Mexico, and observers commented that its purpose was mainly to assuage OPEC since the content of the document was highly conciliatory toward that organization.

Crude oil

The eight-man team that prepared the Iepes report focused mainly on the political issues surrounding commercial policy. Having consulted with present and prospective crude oil customers, the team concluded that Mexico could acquire new customers during 1983–85 who would boost Pemex's crude contract volume to 2.5 million b/d. The team noted, however, that Mexico's placing that amount of oil on the market would surely provoke negative reactions from other oil producers.

The team also considered the physical constraints, production capacity, and reserve levels as factors affecting export volumes. Given the limitations imposed by these factors, the Iepes group concluded that sustained exports beyond 1.5 million b/d (the ceiling proposed for the whole decade by the Energy Program of 1980) were not feasible. They speculated that in response to an increase in world demand, Mexico might be able to increase export levels to 1.8–2 million b/d within 6 months to a year.[2]

The major constraint upon increased crude export volumes, according to the Iepes report, is production capacity: at the presidential inauguration of the crude export facilities at Salina Cruz on 10 April 1982, López Portillo claimed that Mexico's ports could handle 3.75 million b/d of exports. Private industry estimates, meanwhile, credit Mexico's facilities with a sustained export capacity of 2 million b/d.

---

[2]This also supposes that domestic demand for petroleum products stays well below the 8.5% average annual growth rate observed during López Portillo's period. Preliminary data suggest that the combined price increases for gasoline put in force December 1981 and 1982 will reduce demand substantially—as much as 6% negative growth in 1983, according to one unofficial estimate.

With proved reserves of crude oil at 57 billion bbl in 1982, the reserves to production ratio was approximately 57:1, thus permitting Mexico to increase production substantially were there a market. There were days in December 1982 when Pemex was pumping 3 million b/d, a level beyond sustainable capacity. In November 1982 capacity was at 2.95 million b/d and domestic demand was at 1.1 million b/d, leaving a potential exportable surplus of 1.85 million b/d. In fact, Mexico exported 1.73 million b/d in September and 1.72 million b/d in October of 1982.

Increases from the 1.85 million b/d level appear very unlikely during the next few years. According to one banking source, Pemex gave its foreign exchange requirements for 1983 as $2.1 billion, mostly for capital goods. Such an amount, if spent entirely on oil production, would produce approximately 300,000 b/d of new oil—roughly the amount needed to replace field production decline—according to one estimate of Mexico's overall depletion rate on all fields (10% as a national average).

How much oil a given investment in production might buy in Mexico varies with the area chosen for drilling and the corresponding well productivity. In the Bay of Campeche, an average well might produce 15,000–25,000 b/d. According to one estimate, the decline rate of Campeche light oil fields is 5%/year, while heavy oil fields lose 10–12% annually. In contrast, the onshore depletion rate in the Reforma fields was estimated at 10–20% in 1982. The Bermudez field was reported to experience a 21% depletion rate.

The estimate of an overall national average of 10%, assuming continued water injection, would create a replacement requirement of 290,000 b/d in 1983, a requirement that would go up to 350,000 b/d (12%) in the absence of water injection. All of this suggests that Pemex will try to budget production spending to keep an export capacity in the area of 1.8 million b/d for the foreseeable future.

Natural gas exports

With regard to natural gas, the Iepes team offered two main options for Mexico during the decade: *a*) go ahead with the conversion of the nation's industrial plant to gas or *b*) save the gas. The latter would imply suspending development of gas producing wells in order to increase drilling in other areas, such as those that would produce light crude. They commented that the idea in vogue at the beginning of the López Portillo administration—export major amounts of gas to the U.S.—was unrealistic.

On 3 August 1977, Mexico had signed a letter of intention with six U.S. pipeline firms serving 31 states. The agreement, subject to approval by U.S. regulatory agencies, provided for initial Mexican exports of natural gas at a level of 50 MMcfd to increase to 2 bcfd by 1979. The U.S. firms agreed to peg the price to No. 2 fuel oil delivered to New York harbor. The initial price, subject to review every six months, was to be the now-famous $2.60/Mcf.

Mexico, believing that the billion-dollar investment to build the 735-mile, 48-in. gas pipeline from the Cactus field in Chiapas to the U.S. border in Reynosa would have a quick payback period (as short as 200 days) was horrified to learn that the Carter Administration, which they felt was influenced by pro-Alaskan, pro-Canadian lobbies, had decided not to authorize the pipeline consortium to go ahead with the gas import deal. Having made the investment—an overinvestment, to the benefit of private interests, said some—the Mexican president complained loudly that he was left "hanging by a paintbrush." President Carter and his Energy Secretary were excoriated in Mexico, and there was general enthusiasm when Secretary James Schlesinger resigned.[3]

The Iepes team observed that during 1980–82 the industrial as well as the residential sectors were encouraged to increase their consumption of natural gas. Although this policy succeeded in increasing significantly the share of natural gas in total energy consumption, in many cases the gas was used in a highly inefficient manner—especially, they noted, by the petroleum sector itself.

For this reason the Iepes team recommended giving priority to devising gas-saving measures, especially in the energy sector, either through the installation of equipment that would burn residual fuel or through the replacement of existing gas-burning equipment by more efficient installations that would reduce the consumption of gas in power generation and refining.

Concerning the volume of natural gas exported during the 1980s, the Iepes team focused on the pivotal importance of three issues: the advances in converting the industrial plant to natural gas, the level of crude oil production reached, and the reduction of the large gas losses in the petroleum fields of Campeche Sound. In this respect, changes in the structure of internal prices for natural gas as well as residual fuels increase for the medium term the volume of gas available for export and reduce the export sales of residual fuels whose market price is less, in calorific terms, than natural gas.

In addition to the influence of internal considerations, the amount of Mexican gas available for export will also be determined by the capacity of the foreign market (the U.S. market) to absorb it, the policies followed with regard to price fixing, and the export volumes of the closest competitor to Mexican gas exports, Canada. In the 1980s, the Iepes team sees Mexico in an increasingly competitive position with Canadian gas producers and in a good position to maintain current prices in real terms. Maintaining coordinated pricing and export policies with Canada vis-a-vis the U.S. will doubtless be a central theme in Mexico's approach to natural gas exports during the 1980s.

---

[3]For more information on the natural gas affair, see G. Grayson's *Politics of Mexican Oil* (1980) and H. Castillo/Ruis' *Huele a Gas: Los misterios del gasoducto* (1977).

Planning

Reading Mexican discussions of hydrocarbon and nonoil exports, one sometimes gets the impression that in Mexico "exports" means nothing more than freight-forwarding. Sales are left out.

Mexico has tried to approach export sales as a matter of political negotiation. This point of view lay behind Mexico's market-diversification strategy. The theory was that the governments of Mexico-like countries around the world would appreciate and be willing to pay more for non-Middle East sources of crude. The paying-more-for part was initially a subtle matter. Prospective buyers were told that Mexico expected some form of economic reciprocity in exchange for Mexican oil contracts.

Basically Mexico's oil was bartered for so many dollars per barrel plus throw-ins in the areas of technology, investments, financing, and export agreements. Nothing was formalized, however, and country buyers were sometimes left in the dark about the terms of their implied quid pro quo. This certainly was France's situation when, in June 1981, the French government was astounded to learn that, as a direct result of the CFP's decision to suspend third-quarter lifting in Mexico, over a billion dollars of French contracts in Mexico were about to be lost. CFP's liftings were resumed in August, and French contracts—and faces on both sides—were saved.

The CFP affair was nothing compared to the loss of face in the aborted natural gas sales to the U.S. in 1977. A tremendous effort had been made on Mexico's part, in dollars, man-years, and materials, to build a 48-in. pipeline from the Yucatan peninsula to the Texan border. Good industrial reasons supported this project, principally Pemex's problem of having to flare huge volumes of natural gas each day for lack of storage facilities but mainly for lack of a market. Exporting that gas to the U.S. was the logical solution, and one that was technically and commercially feasible from Mexico's point of view.[4] But the point is that Mexico had failed to close the sale, which is always the seller's responsibility, not the buyer's.

Pricing Strategy

Mexico's sales problem in 1977—and this resurfaced in the third and fourth quarters of 1981—was pricing strategy. Throughout this period Mexico had kept company with the price hawks of OPEC who said, in essence, "The price of oil is politically determined. Our price cannot be compared to those in the marketplace when all things [read: 'sales features'] are considered. Accept our price or expect sales blackballing in the future."

This approach to pricing policy worked pretty well under crude-scarcity conditions in the international marketplace. During that time, Pemex and the

---

[4]For the DOE's point of view, see G. Grayson's *Politics of Mexican Oil* (1980).

Mexican government were able to import large quantities of foreign loans on the strength of Pemex's ability to repay through exporting oil at high prices. With the turn from scarcity to surplus conditions, the country is in a bind: these loans have to be repaid ($28 billion in commercial bank debt came due in 1982, and 1983 debt service payments are approximately $14 billion), and the demand for oil is not the comforting certainty so many had supposed.

The apparent conclusion is that the political approach to export pricing may have worked well during 1973–1980 when seemingly all the cards were in the hands of OPEC players. But this pricing strategy is obsolete under surplus oil conditions. It is equally apparent that success in oil industry investments, from a production point of view, does not guarantee success in hydrocarbon exports and has nothing to do with success in nonoil exports.

The Legacy of López Portillo

With regard to the much-discussed Energy Program, the greatest constraint affecting its possible implementation is political. Because Mexico runs on a strict presidential system, there is no practical way for one administration to guarantee that long-term policy objectives expressed in quantitative terms will be followed by its successor. This is why documents like the Industrial Development Plan gave precise formulation up through 1982 but only general indications beyond that date.

The Master Plan for the Petroleum and Petrochemical Industries, the Energy Program, the Industrial Development Plan, and the Global Plan were supposed to mesh nicely and provide the next administration with a rational, unambiguous foundation for domestic and foreign economic policy. The extent to which the letter of the economic planning documents of the López Portillo administration will be adhered to is an open question, but there can be little doubt that the spirit of these documents will guide successive administrations. In truth, these documents are themselves the expression, in quantified terms, of much of the Third-World economic nationalism of the Echeverría period. The López Portillo administration will be remembered in Mexico for having made unprecedented efforts to rationalize, in the context of world-scarcity conditions, a framework for negotiation with present and prospective buyers of Mexico's hydrocarbons. What remains to be seen is the extent to which this framework can be adapted to the unexpected world surplus conditions of the 1980s.

From a completely different point of view, there is reason for concern that the program may succeed but with unanticipated side effects. Having achieved the industrial targets set by the Energy Program and the Industrial and Global Development Plans, Mexico could suffer the economic ossification experienced in the Soviet Union, brought about by centralized economic planning and control.

There are already examples in Mexican industry where the negative effects of hyperplanning may be observed. For example, protestations to the contrary notwithstanding, Mexico's development plan for the fishing industry is on the rocks, at least insofar as the tuna subsector is concerned. A 6-year plan promised to build Mexico's tuna fleet to a class competitive in size and technology to that of the U.S. and Japan. The plan may yet succeed except for one important detail: a market.

In July 1980 Mexico took actions that knowingly would close the U.S. market to Mexico's tuna exports. Mexico insisted that the U.S. respect Mexico's 200-mile economic zone. Tuna, although migratory, are still in Mexican waters, the Mexican government position says, and any fishing in Mexican waters must therefore be by vessels under Mexican flag. This is fine, say Mexican and U.S.-Mexican joint-ventured ship owners, but, having caught my tuna under Mexican flag, where am I going to sell it? One result of this classic Mexican standoff is that U.S. as well as Mexican tuna vessels are seeking to exchange their Mexican flag for others more profitable from the standpoint of markets and sales.

Fishing, like oil, involves a natural resource and is an Article 27 industry. Therefore, political eruptions are always possible. The lesson here, that the laws of the marketplace must be respected and obeyed, is not one that government planners learn easily. One businessperson in Mexico, commenting on this problem, observed that the López Portillo government had planned for everything—except contingency. "But does Mexico need a National Contingency Plan?" he asked ironically.

Intimations of de la Madrid

What made such healthy growth rates of domestic petroleum products sales possible during López Portillo's time was Mexico's policy of requiring Pemex to sell its products at discounted prices. In 1982 shortly after his selection as the PRI's presidential candidate, Miguel de la Madrid Hurtado explained the rationale behind discounting petroleum prices:

> *Nationalism and egalitarian society are together the principles that have oriented and should continue to orient the energy policy of the governments descended from the Mexican Revolution. From these postulates, it follows that the essential aims of the energy sector enterprises [Pemex] are permanently pledged to the support of independent economic development and raising the level of well-being of the population.*[5]

In other words, Pemex does not exist primarily to make a profit, and the condition of the state oil company's income statement takes second place to the

---

[5]Miguel de la Madrid Hurtado, "Energía y desarrollo nacional," *Energeticos* (June 1982), p. 2.

economic condition of the general population. If discounted petroleum product prices help the Mexican people, then prices will be held down by government fiat. Simple enough, and also a politically necessary doctrine.

The magnitude of the subsidies to petroleum consumption in Mexico representing foregone income for Pemex and a strain on its oil production capacity, coupled with the severe financial reverses that hit in 1982, have forced the government to reevaluate its policy on discounts. In the same address from which the above excerpt was taken, de la Madrid attempts an explanation for a change in pricing policies that does not conflict with the nationalistic and egalitarian principles that he had just claimed guide Mexican energy policy.

First, he said, the government does not wish to stimulate wasteful consumption, but neither will it renounce energy policies that promote social well-being by redistributing income to the majority by a price policy that permits greater energy consumption by the general populace. Low energy prices transfer money to the lower classes in cash-equivalent terms.

Second, within that wide context of objectives, de la Madrid called for a revision of prices and subsidies that can still contribute toward the creation of the egalitarian society envisioned by the Mexican Revolution. Prices, according to de la Madrid, "should lead to greater rationalization of consumption and assure that the energy sector enterprises sell their products at an average price no less than their cost of production, promoting in this way the healthy financing of their expansion."

For the first time, the government introduces the concept that the Mexican people "have an obligation to nurture the financial health of their nationalized industries." This statement by de la Madrid, while it may not contradict his earlier pronouncement on the rationale for subsidizing prices, certainly shifts the emphasis of energy pricing policies from support of the individual consumer to support of the economy's ill health, as represented by the penurious condition of Pemex and other state enterprises. Some observers regard this policy shift as a way of putting a good deal of the blame for petroleum product price increases on the previous administration—of which de la Madrid himself was very much a part.

While the economic reasons for lowering subsidies and raising prices on energy products are perfectly understandable, for de la Madrid to ascribe the financial difficulties of Pemex to subsidized prices is disingenuous. It is the government's choice to tax approximately 99% of Pemex's gross profits, in effect forcing Pemex to become a debt-encumbering vehicle for the government. Therefore, to say all Mexicans must look out for the financial health of Pemex and other state-owned companies by giving up subsidized consumption, however rational that decision may be, glosses over the fact that the government's fiscal policies are responsible for Pemex's financial condition—not the subsidies.

What de la Madrid was avoiding saying is that the government has decided that the lesser of two evils is depriving Mexican consumers of the customary level of subsidies to petroleum consumption. The more drastic evil would be to continue the subsidy levels and thus continue depriving Pemex of income that could go toward repaying the nation's foreign debt as well as continue encouraging wasteful domestic consumption of petroleum that would otherwise be freed for the export market. In short, international financial and political pressures on the Mexican government are more onerous than the domestic political consequence of reducing subsidies to petroleum consumption.[6]

One area of contingency is imports. Not all of Pemex's requirements for products, equipment, and supplies can be found in Mexico from local suppliers. One of the jobs of Pemex's officials around the world is to coordinate imports on behalf of Pemex itself as well as on behalf of corporate friends in Mexico who, in turn, provide oil-field services to Pemex.

---

[6]On April 7, 1983, the price of diesel was raised from 10 pesos/liter to 14, leaded gasoline (Nova) from 20 to 24 pesos, and unleaded (Extra) from 30 to 35 pesos. Based on 1982 sales, this increase will generate $2 million additional income per day (300 million pesos).

6

Procurement

Sources of Supply

Pemex purchases petroleum and petrochemical products as well as supplies and equipment. Naturally, as the operator of a nationalized industry, all of the petroleum and petrochemical products purchased represent those that Pemex could not produce itself and so must be imported. Supplies and equipment are bought from both foreign and domestic suppliers.

In Mexico's overall balance of trade, Pemex since 1976 has made a very positive contribution, due of course to crude exports (Fig. 6.1). Nevertheless, Pemex has imported significant levels of petrochemicals and petroleum products. The dollar value of petrochemical imports grew at a 78% annual rate from 1978 to 1980, indicating a serious deficiency in Pemex's petrochemical productive capacity. However, the heavy investments in that sector began to show up in the import figures for 1981. The value of petrochemical imports in 1981 was only 1.2% more than the 1980 value, going from $522.8 million to $523.5 million.

Petroleum product imports followed a similar pattern, with a 29.9% average annual growth from 1978 to 1980. Even more dramatic than the slow-up in petrochemical imports, product imports fell to $159.1 million in 1981, down 34.6% from a 1980 level of $243.2 million. By 1981, two categories of Pemex exports almost exactly equaled the dollar value of Pemex imports of petrochemicals and petroleum products. The value of natural gas exports just slightly exceeded the value of petrochemical imports, and exports of petrochemicals were slightly below the value of petroleum product imports. That left Pemex in the position of being able, in effect, to pay for all of its product and petrochemical imports with its surplus natural gas and petrochemical production.

An important means of reducing the expenditures of currency on imported products has been through the use of *maquila* agreements. Pemex ships crude oil, liquefied gas, and ethylene to refineries and petrochemical plants in Spain, France, Italy, and the U.S. for processing and in return receives products and

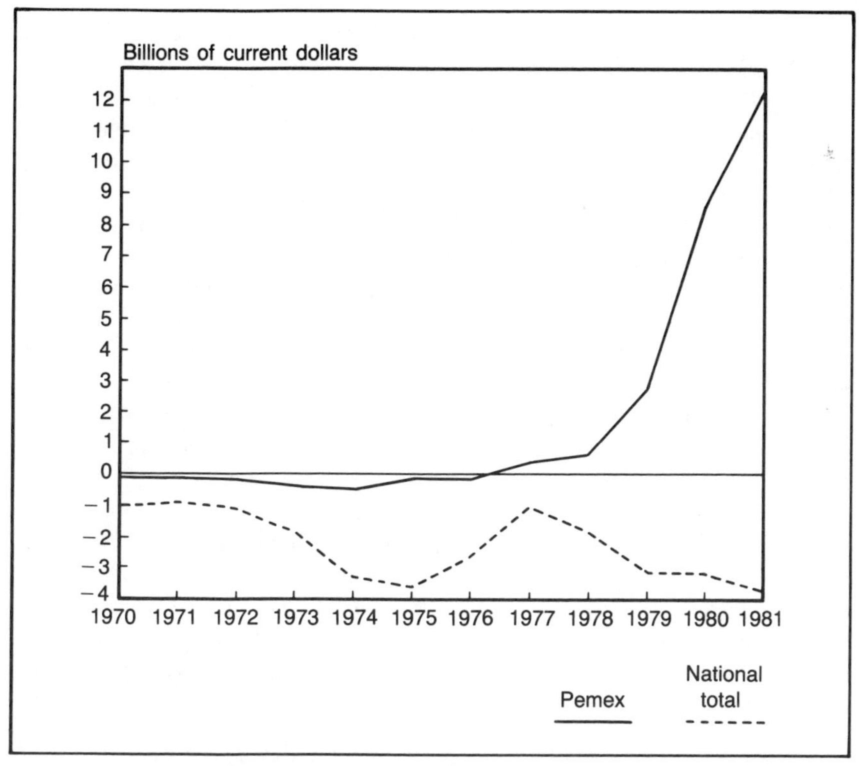

Fig. 6.1 *Pemex leads Mexico's trade 1970–1981 (source: Banco de México)*

petrochemicals. Through this method Pemex saves the expenditure of foreign currency equal to the value of the raw materials that go into the processing of the petrochemicals and petroleum products purchased.

Purchases of supplies and equipment by Pemex, both for operations and capital projects, are made on a preferential basis from local suppliers. In general, Pemex, given equal specifications, will buy the lowest-priced item. However, a local supplier's price may be as much as 15% more than the landed price of the foreign item, and that local supplier may still get the business despite the lower price of the foreign supplier.

Foreign suppliers, who must be registered with both the Secretariat of Commerce and with Pemex before being eligible to sell to Pemex, primarily conclude supply contracts when the product is either not made in Mexico or cannot be provided by the required delivery date.[1]

---

[1]Frederick J. Tower's *Doing Business with Pemex* (U.S. Dept. of Commerce, 1981), reprinted in *Energy Mexico* (U.S.-Mexico Chamber of Commerce, 1981), gives specific information on Pemex offices in the U.S., pricing, billing, payment procedures, and directions for registering as a supplier to Pemex.

New products—a better mousetrap for oil spills, for example—will be field tested in Mexico, usually under the routine auspices of the IMP. In one instance a U.S. oil-spill equipment manufacturer was in a holding pattern for months while testing on his product was being scheduled. Then one day a phone call came from the company's agent in Mexico City saying that there was an oil-spill emergency and could he fly to Mexico to demonstrate the special application of his company's product? "Let's not waste time," he replied, "put me in contact directly with the Pemex people in charge of the problem." The U.S. company absorbed the cost of the flight under promotional expense, and as it turned out the product under test performed magnificently. Pemex managers were appreciative of the attitude and response of the U.S. company; however, over two years later the company finally received a purchase order from Pemex.

The aspiring exporter can get some assistance in paperwork matters from Pemex's overseas offices. In Mexico, thanks to the open-door procurement policy started under de la Madrid, a sales manager need only call Pemex's purchasing office in Mexico City, the telephone number of which was given out in the widely advertised, $1.77 billion material procurement program for 1983.

Measured in current pesos, Mexico provided 51% of Pemex's supplies and equipment needs in 1980 and 56% in 1981. Of foreign sources of supply, only the U.S. lost ground in percentage terms from 1980 to 1981, falling from 37% of total supplies and equipment to 25%. European suppliers picked up 4% more of the market, and Japan 3% (Fig. 6.2).

Effect of Exchange Rate Subsidy

As discussed in the chapter on investment, the overvaluation of the peso relative to the dollar makes it less costly in pesos to buy imported goods than if the exchange rate were based on the parity of purchasing power of the two currencies. The purpose in this section is to examine the effect of an overvalued peso—in effect, an exchange rate subsidy—on two aspects of Pemex's purchase of imported supplies and equipment. First, one must calculate how much those imports actually cost the Mexican economy, not just the nominal value as reported by Pemex. Second, one must look at how these adjusted values change the percentage relationship between foreign and domestic sources of supply.

To calculate the cost to the Mexican economy of Pemex's imports of supplies and equipment, we take the current peso figure reported by Pemex and convert it to dollars at the official exchange rate for the year. That figure represents the number of dollars actually acquired through the expenditures of pesos at the subsidized rate. By converting those dollars back to pesos at that year's parity adjusted exchange rate, we arrive at the "real cost" of the dollars in pesos.[2] Real cost means the amount of pesos that would have had to be spent

[2]The parity-adjusted exchange rate for each year is calculated by taking the 1960 exchange rate, 12½:1, multiplying it by the value of that year's GDP deflator with 1960 as the base year, and dividing the product by the U.S. deflator of the same base year.

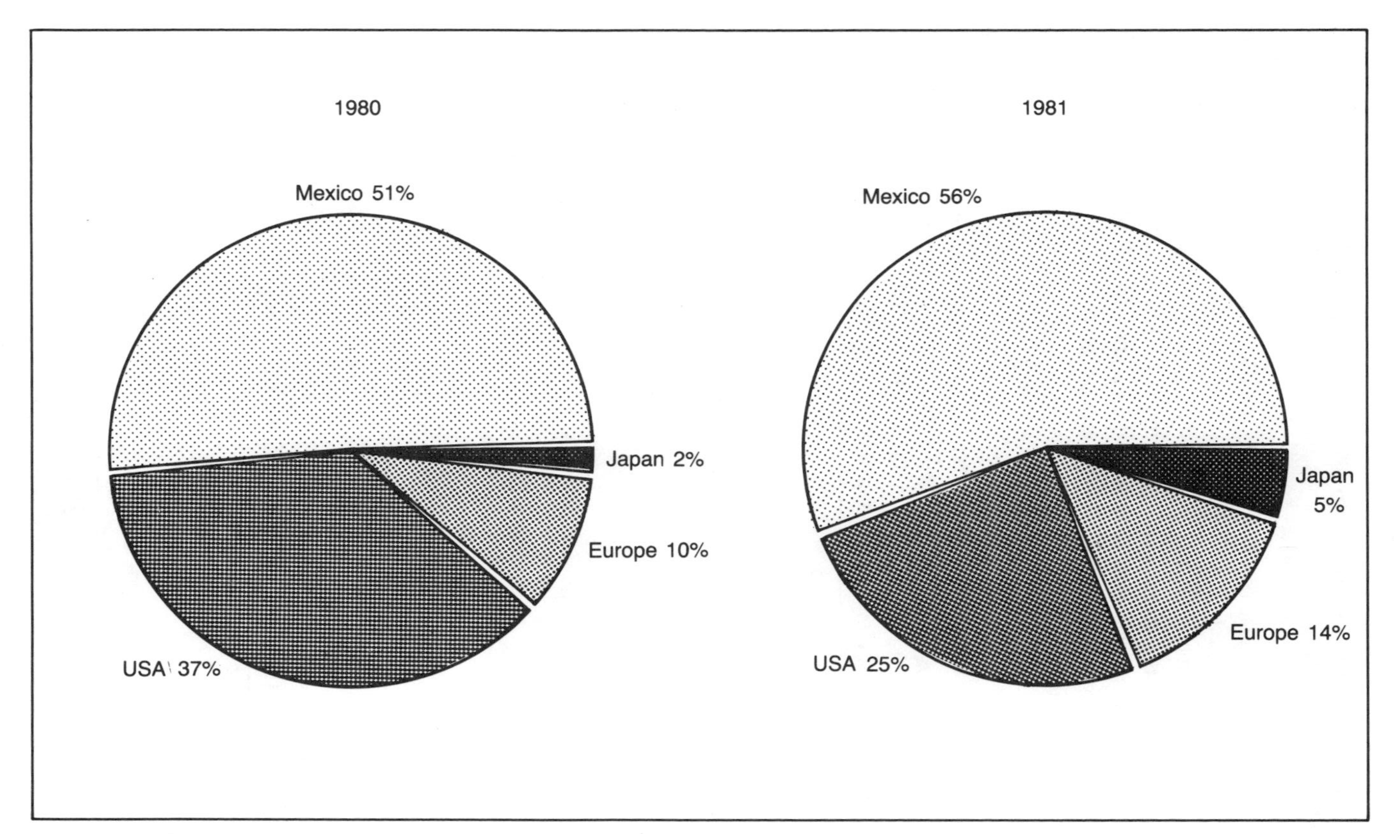

Fig. 6.2 *Pemex source of supplies and equipment (source: Pemex)*

by Pemex if the peso's value against the dollar had been adjusted for differences in inflation in the two economies—in other words, adjusted for differences in purchasing power.

That cost still must be absorbed by the Mexican economy as a whole in terms of the overvaluation's hindrance to Mexican exports, including tourism, and its encouragement of higher levels of imports, with attendant negative effects on Mexico's balance of payments and foreign reserves (Fig. 6.3). Converting the real peso cost back to dollars at the official rate and subtracting from it the nominal dollar value of imports gives the dollar value of the exchange rate subsidy (Table 6.1).

This method of evaluating Pemex's foreign purchases does not yield an exact measure of the effects of an overvalued peso since Pemex also buys from non-U.S. sources in other foreign currencies. However, because the U.S. is the dominant foreign supplier, providing 75% of imported supplies and equipment in 1980 and 57% in 1981, this method indicates the approximate effect of overvaluing the peso. The peso has also been consistently overvalued against the

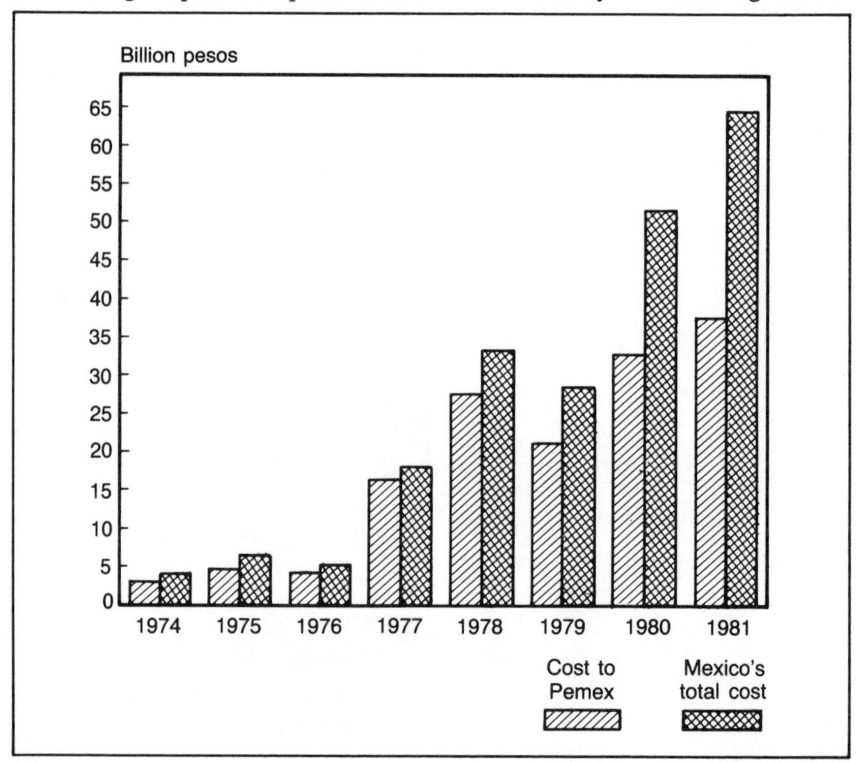

Fig. 6.3 *How Mexico subsidizes Pemex merchandise imports (effect of exchange-rate subsidy), (source: Baker & Associates)*

Table 6.1
Exchange rate subsidizes imports of Pemex's supplies and equipment

	1974	*1975*	*1976*	*1977*	*1978*	*1979*	*1980*	*1981*
				(Millions of current pesos)				
Total procurement								
Local	5,091	8,432	5,697	18,230	20,635	23,569	35,500	48,335
Imports	3,169	4,751	4,176	16,329	27,615	21,209	32,768	37,596
Total	8,260	13,183	9,873	34,559	48,250	44,778	68,268	85,931
				(Percent)				
Local	62	64	58	53	43	53	52	56
Imports	38	36	42	47	57	47	48	44
				(Millions of equiv. dollars)				
Imports only								
Nominal	254	380	270	723	1,213	930	1,428	1,534
Ex-rate subsidy	75	146	80	77	252	318	821	1,090
Revised total	329	526	351	800	1,465	1,248	2,249	2,624
Subsidy's effect on import cost (%)	29.68	38.48	29.73	10.67	20.82	34.20	57.52	71.07
				(Millions of parity-adjusted current pesos)				
Adjusted Total								
Local	5,091	8,432	5,697	18,230	20,635	23,569	35,500	48,335
Imports	4,110	6,579	5,417	18,072	33,364	28,462	51,615	64,317
Total	9,201	15,011	11,114	36,302	53,999	52,031	87,115	112,652
				(Percent)				
Local	55	56	51	50	38	45	41	43
Imports (Parity)	45	44	49	50	62	55	59	57
Effect of Parity Adjustment on imports (%)	6.30	7.79	6.45	2.53	4.55	7.34	11.25	13.34

Source: Procurement figures from Pemex; adjusted by Baker & Associates

Japanese yen and most European currencies, so the impact of overvaluation relative to the currencies of Pemex's non-U.S. suppliers is in the same direction and general magnitude as in the case of dollar-denominated imports.

Adding the subsidy value of the overvalued peso to Pemex's peso expenditures for imported supplies and equipment increased the percentage share of imported goods in Pemex's purchases. Local suppliers provided more than 50% of Pemex requirements in every year but one between 1974 and 1981 but provided more than 50% in only three of those eight years after adjusting for the exchange rate subsidy. The parity adjustment increased the share of imports in Pemex's purchase of supplies and equipment by more than 11% in 1980 and by more than 13% in 1981 (Table 6.1).

Imports of Capital Goods

The Mexican oil industry's capital goods requirements between 1983 and 1988 will total approximately $20 billion, according to Guillermo de la Mora, head of the planning division for refining of the Mexican Petroleum Institute. At least one-half of that figure, $10 billion, will have to be imported.

For U.S. suppliers that means a market in the neighborhood of $5.7 billion, provided they maintain their 1981 share of Pemex's imported goods. The breakdown of Pemex's capital goods requirements for 1982–1988 are as follows:

- $8.0 billion: drilling and production
- $3.6 billion: refining and petrochemicals
- $8.4 billion: transportation and distribution

If crude export levels rise to 2 million b/d, according to an executive for a Mexican metal goods supplier to Pemex, the capital goods needed by Pemex could reach $30 billion at 1981 prices. In addition to the extra capital goods required to sustain exports at 2 million b/d, Pemex historically has underestimated its capital goods expenditures so that a figure higher than the $20 billion is conceivable. If the $30 billion level is reached, that would push the market for U.S. suppliers to $8.6 billion.

Procurement Intermediaries

As in other areas of Mexico's oil industry, there is a normal distribution of procurement consultants, go-betweens, and traders, some of whom exist as bona fide commercial entities while others exist as concession-holders whose right to procure goods and services for Pemex is due to political or family connections.

Inevitably, as in all LDCs, commercial rights granted by the state are awarded as concessions or permits, as distinct, for example, from an auction where appreciation by the concessionee is likely to be expressed in monetary terms.

Human nature requires that a certain number of concession givers and consultants go into business charging a cut, or markup, on the commercial value of the concessions. This state of affairs seems to have been the cause that led to the 1982 charge and conviction of a U.S. oil-field equipment executive under the Foreign Corrupt Practices Act of 1977. Evidence introduced in court showed that two Pemex officials allegedly charged 5%, or $9.9 million, of the purchase order value of compression equipment contracts. This was the first instance in which a conviction under the 1977 act was obtained. The two Pemex officials reportedly left Mexico—in haste.[3]

With regard to the procurement of oil-field services from abroad, only a few generalizations can be made. One U.S. company that offered to computerize Pemex well logs was turned down, in part, because Pemex did not want foreigners seeing the primary data. As is generally the case in LDCs, Pemex will choose name-brand, blue-ribbon companies over smaller, less-known (but perhaps better) firms. This tendency could be reversed, depending on the outcome of the multimillion-dollar suit filed in Houston in December 1982 by Pemex against three major U.S. oil-field companies for, in the words of Pemex's brief, "conspiracy, collusion, and for having prepared fraudulent estimates and offers."[4]

In general, besides following normal legal and administrative steps for doing business with a Mexican government agency, a long-term marketing effort in Pemex and the IMP is mandatory. Private companies in Mexico's oil sector (whose names are in PennWell's annual *Latin American Petroleum Directory*) may be of interest as agents, partners, or customers.

Unfortunately, the checks written for Pemex's imports are not distributed in the bookkeeping sense of the term by investment activity. No one knows, from published sources at least, how much Pemex and its contractors spend on imports, for example, in the Gulf of Campeche. No one therefore knows, on a per-barrel basis, how many dollars are needed to get a barrel of oil from an offshore well there.

---

[3]In early April 1983, responding to charges of corruption within Pemex, the attorney general of Mexico asked for an immediate audit of Pemex's office of administrative management and of the subdirectorate of supply and warehousing to discover by how much Pemex was defrauded by former officials. The controller general reported that audits had already started on the supply management office and the subdirectorate of warehousing.

[4]*Mexican News Synopsis* (December 20, 1982), p. 10.

7

Investment costs and returns

Unit investment costs and profits in Mexico's oil sector are two of the most deeply buried bones in Pemex's backyard. Only the scent of such costs and profits lingers.

Drilling for unit data

The overall profitability of the oil sector in Mexico has been calculated in Mexican official publications as the "value added" by Pemex. The value added by Pemex is total output times market price less intermediate consumption, i.e., consumption of goods and services that Pemex buys from third parties.

This method of analysis gives some idea of the profitability of the company but does not touch on the profitability of its investments in oil production or in any other activity. The method, as practiced with Mexican data, is likely to understate Pemex's operational profits, owing principally to the effect of domestic price subsidies on Pemex's domestic sales. As Jaime Corredor noted in 1980 while on the staff of the Office of the President of Mexico, Pemex's sales in 1979 would have been $22.986 billion at world prices, not the reported $7.286 million. Another factor affecting the calculation of Pemex's value added is the subsidy given to Mexican importers, including Pemex, during most of the López Portillo period.

To deal with the question of the profitability of Mexico's greater petroleum industry, including by that term the several hundred Mexican oil-field supply and service companies, is to invite the computer to crash owing to too many NAs (not available) in place of data. Pemex, needless to say, does not choose to publish data on the profitability (or unprofitability) of its producing districts.

Yet some corner of the financial veil in front of Mexico's investment program in crude production can be lifted with generally available data. One question is the investment cost per barrel of oil in any given year. Analysis is almost defeated, owing to the absence of investment data by producing region. Working with investment data in hydrocarbon production for the country as a whole invites the criticism that not a single field has a cost structure equal to aggregate new investment, *I*, divided by aggregate new output, *q*. Take the figures for average hydrocarbon production/well—a mere 711 b/d in 1981. That's an increase from a national average of 201 b/d per well in 1970, but it still understates potential in some regions.

For example, if 67 wells in the Bay of Campeche produced 1.082 million b/d in 1981, the average output per well of crude was 16,149 b/d. The 30 August 1982 issue of *Oil & Gas Journal* reported that in May 1982, 47 Campeche wells were producing 1.326 million b/d, an average of 28,212 b/d per well. Thus, national average data are almost meaningless for interpreting levels of cost and profitability in a given producing region. The Chicontepec region with its reported 17 billion bbl of recoverable reserves is rumored to have extraction costs so high that production efforts are tantamount to a welfare project. At the other extreme, extraction costs in relation to output in the Reforma and Campeche regions make for good profits.

One exercise would be to take the cost of importing and setting up an offshore platform in the Campeche area and assume four to six wells are producing, each with an average output of 20,000 b/d. With five wells, total output of the platform would be 100,000 b/d, which would mean that, at an average export price of $25/bbl, a platform cost of $30 million could be paid for in 12 days of operations.

Theoretical issues

As Professor M.A. Adelman writes, "a price is assumed and multiplied by the declining production profile, to yield an income stream; and each installment of this stream is discounted by such a rate as just to equate the sum of the present value of the forthcoming barrels to the expenditures necessary to install the producing capacity."[1]

The danger in this procedure, as Adelman points out, is that long-term crude price movements cannot be calculated with any degree of confidence, thereby weakening any analysis dependent on the present value of future revenues. Adelman proposes taking a company's cost of capital as the known and solving for the future price of oil, i.e., solving for the various price levels that would *a*) cover current expenses, including management services, were the price of capital zero, *b*) cover current expenses plus the cost of capital (or opportunity cost), and *c*) cover operations and capital costs, with money left

---

[1]M.A. Adelman, *The World Petroleum Market* (Baltimore: Johns Hopkins University Press, 1973), p. 49.

over as profits. As future production declines, the capital cost per unit increases but, as Adelman points out, a production project must be evaluated as a single whole. Capital costs per unit, in other words, will change with time, but they must be reduced to present values to permit an investment decision by management.

While oil production investment analysis has its controversial aspects, such as the degree to which a company should adhere to a net present-value approach to comparing alternative investments, some illustrative analysis of Pemex oil production investments may be undertaken.

After all, a \$3,000–\$5,000 investment/barrel of initial output represents a formidable market for oil-field equipment and services firms, Mexican and foreign. So market size is the other side of the coin of industry profit.

For the moment a distinction can be made between the peso profits and dollar profits of Mexico's oil industry. One approach to analyzing foreign-exchange profits of Mexican oil investments is to assume a long-range export price, multiply that price times the total output in barrels to get total revenue. If all dollar-denominated expenses are subtracted from total revenue, the result will be a simple idea of the foreign-exchange earning potential of the investment.

This approach fails, however, for the same reason Adelman gave earlier: long-range price unpredictability. After the oil price shocks of 1973, 1979, and 1983, no one believes in long-range crude export prices. So the exercise just given, of counting dollar-denominated chickens, may turn out to be accurate, but, like the tree falling in the forest, no one will know about it, at least not in advance. Mexico, like everyone else, including Pemex's bankers, had invested a good deal in the expectation of steady, long-term price increases for crude oil. No one in Mexico wants a repeat of the hangover of 1982, so any analysis predicated on crystal-balling crude export prices will likely be thrown out.

But the point shouldn't be pressed to an extreme. Suppose Mexico's average export price were to stay, in real terms, at \$26/bbl for 20 years. How much money is there to be made? Simply that.

Cost accounting for Pemex's oil investments

As in any oil and gas deal, there is a cost-accounting maze through which one's pencil must pass. Since a good deal is to be learned in the exercise, it's worth going through. Imagine a group of limited partners (the Mexican people) who have title to speculative oil properties and who want to invest some of their money in one of two projects under consideration, both in the Gulf of Campeche. One of their questions is which one? Another is, Since the work is going to be carried out and managed by others, how do we know we're not being cheated?

The limited partners contract the services of a general partner (the Mexican government) who, in turn, turns the project over to a semi-integrated oil concern (Pemex). The oil company assumes responsibility for project management but farms out drilling and production to a Mexican subcontractor like Permargo or assigns it to an in-house company, its primary production unit.

The ensuing accounting drama is one in which the drilling company charges the oil company for time and materials, plus indirect overhead charges and a management fee. The oil company charges the general partner for direct costs, indirect costs, and a project management fee as well. The general partner finally charges the account of the limited partners for all costs up to this point plus a management fee of its own.

What are the costs, and who keeps track of them?

The drilling company has responsibility for the raw investment cost: all costs directly associated with producing the initial daily output of oil. Also, indirect costs, beyond those of normal overhead, are associated with getting the initial oil production onstream and with maintaining field operations. One of these costs will be the cost of procuring imported supplies, capital goods, and services. In Mexico the drilling company often looks to Pemex to be the importer of record for such supplies and equipment. When dollars became scarce in 1982, this role of Pemex became especially important for Mexican arms-length contractors. Pemex's profit-margin on these vital middleman transactions is not reported as such, although the cost of acquiring "merchandise for resale" is given in the annual report.

Assuming that the cost of such Pemex services is passed on to the drilling company, the only remaining cost that Pemex must bear is that of project management, some portion of which might be subcontracted to the IMP. Pemex's March 18th statement to the general partner, therefore, includes all invoices from the drilling company and its project management fee as well. The report also lists an accounting of the revenues collected by Pemex as the commercial collections agency representing the interests of the limited partners. Given Pemex's contractual relationship to the general partner, not a dime of the money collected from the sale of crude oil belongs to Pemex itself. Having purchased crude from the drilling company at the transfer price set by the general partner, Pemex can claim that its refineries contributed value of their own by producing petroleum products.

The general partner's September 1 report to the limited partners includes all of Pemex's costs plus the cost of leveraging the money the limited partners put up. In practice, Pemex reports interest payments on capital projects, but the fact that Pemex is the legal borrower of dollar-denominated loans is a banking fiction. Pemex does not have title to the oil it produces, for which reason its ability to repay foreign loans is logically and contractually unrelated to the export price of Mexico's crude oil. Pemex's ability to repay foreign loans is a function, first, of its negotiated management fee with the general partner and, second, a convertibility of pesos into dollars.

Evaluating alternative oil production projects

How much money is going to be left for the limited partners representing the private sector of the Mexican economy? The partners understand that the initial daily production will decline over time. They, unlike everyone else in the equation, must calculate their money tie-up cost because they are putting their money in this speculative venture in lieu of putting it in U.S. Treasury bills. The amount of the interest income they forego can be regarded as the tie-up cost. The rate on such 99% secure investments used to be in the mid-1960s around 5%. In May 1981 U.S. Treasury security yields for 10-year constant maturities reached 14.1%. Therefore, for the investors some of the projected profit must be reduced by discounting it at the rate used as the cost of tying up their money.

Suppose that the investors are evaluating two possible deals. Which is the better investment? Knowing that their profit will be the price of oil less all costs, including the cost of tying up their money, they choose the investment with the lowest total investment cost/barrel expressed in terms of the purchasing power of today's money. The expected commercial life of the two deals may or may not be equal, and they may differ in their expected decline, or depletion, rates. The money tie-up cost will be equal, in terms of the rate, for both deals, but the total amount will differ according to the amount of money to be tied up in a given deal.

For each deal there will be a time, t, representing its commercial life, an average annual depletion rate, α, and a money tie-up rate, r. The two rates can be combined into a single discount rate that is to be applied to output over the period, starting with the initial daily output. Having adjusted for these factors, two deals will differ in terms of their FTE (full-time equivalent) factor, which is the length of time that an oil project may be regarded as producing at the initial daily output rate. In other words, instead of showing how output is reduced over time, the idea is to calculate how time is reduced with the initial output held constant.

Project A vs Project B

Project A is for 25 years with a 10% decline rate and a 10% money tie-up cost. Project B is for 20 years, has a 15% decline rate, and has the same tie-up. Assume that the dollar investment required for each project were $1 billion, to include all drilling company and Pemex project management costs and that the initial output were 300,000 b/d for each project. Project A's FTE life is 5.43 years or 1,982 days, while Project B's is 4.43 years or 1,617 days.

The per-barrel investment needed for the initial daily output of the two projects is the same ($1 billion/300,000 b/d). However, the two projects will differ when that figure is scaled by their respective FTE production factor. Project A has a raw investment and capital tie-up cost of $1.68/bbl, while Project B's is $2.06/bbl.

The decision for the Mexican investor group favors Project A, whose cost/barrel in today's money is 18% less than the comparable cost of Project B. An investment in Project A means 38¢/bbl in costs avoided when compared to an investment in Project B. These 38¢ do not necessarily translate into profits, however, since they may be needed for higher oil-field operation and maintenance costs, which might be lower in Project B by 38¢/bbl. Further, the risk in Project A, as perceived by institutional lenders, may be higher, implying a higher cost for leveraging the up-front money of the investors. Even if the sum of these costs on a per-barrel basis were equal to the initial investment and capital tie-up cost, it would still mean that investors were looking at a potential profit, again, in today's money, of $22.64/bbl in Project A and $21.88/bbl in Project B.

Regressional analysis using national data

There is a temptation to unleash the full force of statistical regression theory on what Mexican cost and production data may be available. In this way one may come across crumbs that lead back to an enchanted, explanatory model of epic proportions. With an ultra-small sample of a dozen annual observations covering aggregated national costs and output, there is a fallacy. For even an alchemist needs a little gold to make more, and it's not clear that national, annual data constitute any gold at all. Quarterly data by zone? Yes. Annual data for the whole country? Doubtful.

But beggars, especially statistician types, ought to be grateful for the data they do have; wringing their hands over the data they would like to have is not what they're paid for. Data are available for Mexican oil production by zone, and it would be worth someone's time to experiment with annual D&P investment data by prorating investment costs by zone. The statistical problem is to solve for the investment cost coefficients by zone, given total costs, total output, and partial output by zone. Parallel problems requiring solutions are a) depletion rates and b) distributed lag times between investments and output, by zone.

The question of how much new investment is required for a given quantity of new output also can be attacked broadside from published data. Knowing Pemex production data for crude and total hydrocarbons for a period, say 1974–1981, and knowing total Pemex investments in development drilling and production during the same period, it would be possible to come to conclusions about year-to-year costs of new production (Table 7.1).

Table 7.1
What Mexico invests in production

Year	*Total hydrocarbon development & production investment (Millions 1981$)*	*Total hydrocarbon output (including natural gas @ 6 Mcf/BOE) (MMbbl oil equivalent)*
1974	505	354
1975	574	413
1976	1,094	444
1977	894	508
1978	2,030	627
1979	2,468	749
1980	2,959	971
1981	3,660	1,150

Source: Pemex; adjustments by Baker & Associates, Table 3.5

New output will be proportionate to effort (in dollars) in relation to the difficulty of extraction and well productivity. Where, as in Pemex's case, new output is not given at all and is a quantity to be derived based on an inferred depletion rate, the matter is more problematic. The only true independent variables are the amounts of monies spent in hydrocarbon production for the nation as a whole by Pemex in a given year.

Start with a linear equation showing net new output, q, as determined by the sum of a constant, A, and a well productivity factor, α, times the money invested in production (Fig. 7.1, line 1). In any given year, the oil produced from previous years' investments would be equal to last year's q less the applicable depletion. It follows that gross output in a given year is last year's output less depletion plus new (or net) output.

In general, unit investment costs vary inversely with field and well productivity but eventually rise as a reservoir is used up and extraction becomes more difficult. This means that well productivity, α, varies with time. For not being able to measure well productivity directly, a proxy would be to take the inverse of the sum of all previous output as a factor by which the efficiency of new investments should be reduced. In Mexico's case, with so much new oil discovered during the late 1970s, the efficiency of investment capital is likely to remain even, or possibly improve, at least for a while.

All of this assumes that investment in a given year produces new, or net, output in that year; however, in Mexico, as elsewhere, the lag time between investment and new output may vary from 6 months to up to 2 years. A lag time of 2 years would mean that the new output of 1980 should be regarded as having been caused by the 1978 investment in production.

To calculate the average cost of new production, one simple method would be to take the net production of a given year divided by its corresponding

Output (Q) is proportional to *investment* (I), the productivity of which is affected by geological, management, and technological factors (together represented by α):

(1) $Q_o = A + \alpha I_o$
$Q_o =$ first year's output
$A =$ constant term
$\alpha =$ current well (or field) productivity factor
$I_o =$ investment in development drilling and production (D&P) in year *o*

Since oil flow from a well or field normally continues over many years, the calculation of the contribution of prior years' D&P investment to this year's output is last year's gross output less depletion:

(2) $Q_{prior} = Q_{(t-1)} \times (1 - d)$
$t =$ time period under analysis
$d =$ annual reservoir depletion rate

(3) $Q_{gross(t)} = Q_{prior} + Q_{new(t)}$

Well productivity declines as well capacity is used up but rises with increases of management or technology efficiency:

(4) $\alpha = \alpha(t)$
$\alpha(t) = a + b\,[(Q_{t-1} + Q_{t-2} \ldots Q_{t-n})^{-g}]$
$n =$ years of oil flow
$\left(\sum_{j=t-1}^{j=n} Q\right)^{-g} =$ proxy for used-up well capacity

New output (Q_n) is caused by D&P investments made during the period whose distance from the present is the lag time (ℓ) between investment and production:

(6) $Q_{new(t)} = A + \alpha I_{(t-\ell)}$

Hence, when data are annual figures and the lag time is less than one year:

(7) $Q_{new(t)} = A + \alpha I_{(t)}$

Average unit cost (AC) of new oil is Q_{new} matched to its lagged D&P investment:

(8) $AC = \dfrac{I_{(t-\ell)}}{Q_{new(t)}}$

Marginal unit cost (MC) of new oil is the cost of an additional barrel of output and is measured by scaling change (Δ) in investment from one period to the next by the corresponding change in new output:

(9) $MC = \dfrac{\Delta I}{\Delta Q}$

Where the lag between investment and output is greater than one year:

(10) $MC = \dfrac{\Delta I_{(t-\ell)}}{\Delta Q_{net}(t)}$

Fig. 7.1 *Modeling Pemex's investment costs (source: Baker & Associates)*

investment. This figure would be the average cost of new production, which could be taken as tantamount to marginal production.

Strictly speaking, it is theoretically possible to calculate marginal costs, the average cost of new production of one year less the average cost of new production from the year before. Given data for production lag, the foreign-exchange component of production costs, and applicable inflation rates for the U.S. and Mexican petroleum industries and showing total hydrocarbon output, one can take a crack at investments in production in 1981 dollars (Table 7.2). Using a lag time of 2 years, average costs of new production fall out.

Table 7.2
Average cost of new Mexican hydrocarbon production

Year	*Cost 1981$/bbl*
1974	2.59
1975	2.36
1976	4.00
1977	2.63
1978	4.95
1979	4.93
1980	4.50
1981	4.67

Source: Baker & Associates

These figures suggest that new production capacity costs stayed in the area of $2.50/bbl (1981 dollars) from 1974–1976 and rose during 1977–1981 to less than $5/bbl. The jump in 1976's cost is likely to be due to some distortion from the devaluation of that year. The average cost of new production rose substantially in 1978, leveling off for the rest of the period. This period corresponds to the Gulf of Campeche production coming onstream.

The intuitive conclusion to be drawn is that while Campeche production capacity costs were higher than onshore costs, well productivity was so much higher that the return over the full life of the well amply repaid the additional cost.

Are investor profits in dollars or pesos?

Nothing has been said to this point about the desires of the Mexican investors with regard to the currency denomination of their profits. In the bygone days of the fixed and dirty-floated exchange rates, it made no difference. Profits in pesos were easily converted into cheap dollars. When dollars are in scarce supply, profits in pesos are regarded in Mexico as less tangible than profits in dollars.

Two questions are being raised: Isn't it the case that the investor group, knowing that petroleum products in Mexico are sold below international prices, would want to see its investment earmarked for crude oil produced for export? In other words, Isn't production for oil exports basically more profitable than the production of oil for the domestic market? The second question is, what is the basic motive of the general partner? Is the general partner essentially trying to put dollars in the hands of his clients, or is he trying to provide them with units of equivalent purchasing power via subsidized energy products?

The answer to the first question is no. To see that this is so, the mind's eye must recall the business of the drilling contractor: for him it makes not a farthing's difference whether 300,000 b/d go to a Pemex refinery or to a Shell tanker. The economics of his business, oil production, are unmoved by the question of who ends up with the oil. The refined products subsequently produced from that oil are sold at prices that are higher at Shell distributors than at Pemex distributors, but that is a problem for downstreamers to figure out. The driller's only business problem is to produce oil in the most efficient manner.

This answer bears on the answer to the second question, which asks about the inner-most motive of the general partner. With $80–90 billion of dollar-denominated debt, the purpose of oil production would seem to be that of generating foreign exchange. Pemex, in other words, exists as a dollar-cash cow for the investors. If export volume is to be limited by policy decision to the vicinity of 1.5 million b/d, then where are greater dollars to come from, given that the export price of Mexican crude oil is determined by the market, not by the president of Mexico. Practically the only place left is downstream, in the product marketing end of the oil business. Pemex refineries and petrochemical plants should therefore try to export as much residual fuel, aviation gasoline, and ammonia as they can get away with, without stepping on the toes of Venezuela or the U.S. International Trade Commission. At the same time the general partner should try to find ways to lower production costs, especially those in dollars.

In the short term there may be only indirect methods available. One is to lower the capital-leveraging (interest) expense by generating a greater net cash flow at home to pay for peso-denominated production costs and management fees. The question is definitely not one of raising energy prices in the domestic market, the question is merely that of reducing the automatic rebates.

There is a third area where the general partner might earn dollars: by getting Pemex into the oil consulting, exploration, and project-management businesses outside of Mexico. He might propose doing this on his own or in concert with others. Presumably, this was the idea behind Pemex's investment in the Spanish refinery belonging to the company Petronor. Pemex's best hidden investments with dollar-earnings capability are those joint ventures it has made in Mexico. Pemex's annual report shows that the company has a 60% stake in Compania

Méxicana de Exploraciones, S.A., the commercial activities of which included carrying out seismic studies in Costa Rica in 1982. Neither the dollar earnings of this company nor the identity of its minority shareholders is, however, reported. Besides its political overtones, there was at least the smell of dollar earnings in the idea of forming a multinational oil concern in which Mexico, Venezuela, and Brazil were to participate.

Conclusion

Evaluating Pemex hydrocarbon production costs is very difficult and inexact. To Pemex procurement officials under López Portillo, it was a) Pemex's direct labor costs plus supplies and equipment for development drilling and production, b) the direct and indirect costs plus fees of Pemex contractors, and c) Pemex's indirect costs, including the cost of procuring capital goods from abroad for Mexican oil-field contractors and their subs. For Pemex it made little difference whether such costs were in dollars or pesos, since dollars and imports were cheap.

For financial managers in the Mexican treasury department or in the ministry of planning and budgeting, however, the story was somewhat different. It made a great deal of difference whether costs were in pesos or dollars, since imports were consciously being subsidized. For these managers their counterparts in Pemex were in a different world. These managers knew that they were picking up as much as half the tab for Pemex's overseas shopping sprees. Pemex, in short, was buying assets, including real estate abroad (de rigueur under López Portillo), partly with play money.

It's probably true that no one in Pemex's procurement office knew the first thing about the currency overvaluation policy of the government, and it's probably equally true that no one outside of Pemex was allowed to get within 100 yards of procurement data that might correlate oil output with requisitions for goods and services, imported and national. The audit of Pemex's accounts ordered by de la Madrid in 1983 is evidence enough that the left and right hands were far apart.

But with the collapse of the rose-colored world of easy dollars, Pemex and para-Pemex managers had reason to come to a consensus about oil production costs, onshore and offshore. The consensus, at the beginning at least, is likely to be that no one knows. For reasons that Mexico's attorney general started to investigate in 1983, cost/barrel data were not maintained and detailed cost/barrel capital accounting was not undertaken with any seriousness, if at all.

To talk intelligently about hydrocarbon production costs is, one presumes, to know something about historical costs of individual wells or of specific fields. It is to know, moreover, something about the relative costs of oil and gas wells. To admit, therefore, that the only data one has seen cover all producing areas in the aggregate, without regard to field or the difference between oil and gas, is to admit that one has seen very little indeed.

Yet it's questionable if anyone knows more than this very little. It is important to try to sort out what this little bit of knowledge might consist in, especially since Uncle Sam through the Exim Bank in Washington is picking up the bill for Pemex's imports by providing the cash for credit purchases. Although Washington has the left-vs-right hand problem as much as Mexico City, the question of the foreign-exchange component of total crude production costs now becomes an issue of concern for both governments.

The basic fact is that oil production in Mexico is dependent on imports, i.e., on a continuing availability of foreign exchange to buy capital goods and services. Where dollars for oil production-related imports are borrowed, the cost of production necessarily entails the cost of that capital expressed as interest payments. So any proposed approach to oil production costs must account for a) dollar import costs, b) dollar capital interest costs, and c) peso costs for labor, goods, and services, including financial.

All of these costs subtracted from revenues generated yield profits. However, to understand the profitability of oil production in Mexico as a source of net foreign exchange is one question; to understand aggregate profitability in pesos or peso-equivalents is another. The economics of total profitablity is confounded by the issue of domestic price subsidies. As argued in Chapter 2, the only clean way to place a value on Pemex's output of crude, gas, and products is to use international prices. Differences between actual prices charged by Pemex and international prices must be accounted for as either a) subsidies or transfer payments or b) payments by the government for services such as regional development/industrialization. If this argument is accepted, all of Pemex's reported figures for domestic sales must be revised upward. When this is accomplished, total domestic sales in units of equivalent pesos can be discussed. As things stand, Pemex's reported domestic sales figures are hopelessly contaminated by the effects of an entirely alien accounting system that governs energy subsidies.

The measure of the soundness of Mexico's investments in oil production should, one would suppose, show up on Pemex's bottom line. However, this supposition turns out to be quite false. And while the comment is often heard that Pemex's accounting system is perfectly inscrutable, this remark turns out to be not quite just.

8

Financial reporting in the oil sector

Financial reporting in Mexico's oil sector is decentralized, and there are no consolidated statements—not even for the state-owned portion. Pemex is really like a group of companies, each one dedicated to different parts of the oil business: exploration, drilling, refining, jobbing, and export management.

All of this is done under contract with the government. But Pemex the petroleum operation is only part (although no doubt the largest part) of the Mexican oil industry. There is also Pemex the public-works/social-welfare agency of the government. In this capacity Pemex spends more on education, medical services, and worker housing than it receives in net profits. In 1981 the company reported its social service expenditures as 5.7 billion pesos, while its net profit was 400 million pesos.

Such expenditures should, however, be regarded as *de facto* taxes on the oil company, and its income statement should be adjusted accordingly for a measure of oil industry performance.

Engineering R&D is a separate animal, located mainly in the Instituto Mexicano de Petroleo (IMP), usually translated as the Mexican Petroleum Institute, echoing the name—but not the functions—of its American counterpart. The IMP assumes major responsibilities for technical training in the industry, but it is also the government's center of engineering and product development research. The IMP has its own budget, staff, and facilities, but its major "client" is Pemex.

Policy research and development, meanwhile, are most likely decentralized outside Los Pinos (the Mexican White House). The government industry's principal think tank is in Mexico City, on Rio Rhin 22, where the offices of the

Direccion de Energia are located. This is the Energy Section of the Ministry of Energy, Mines and Para-State Industry, known in López Portillo's time as the Ministry of National Patrimony and Industrial Development. Other research centers on oil policy matters are located in Iepes (PRI), the Colegio de Mexico, the National University (UNAM), and in the section of special advisers at Los Pinos as well.

In the U.S., these activities would be incorporated and paid for by the revenues of a Shell or an Exxon. In Mexico's case, Pemex is the principal source of revenues for the industry. The IMP also invoices its clients, but 95% of its revenues come from Pemex and only 5% come from private industry. Pemex pays its own oil-field and administrative costs and then gets roughly 99% of its operating profits taken away in taxes, which in turn pay the salaries and overhead of the other government employees in one way or another assigned to the oil sector. If the cost of these general, administrative, and academic research expenses represents 6% of total income, then oil sector revenues of $15 billion would imply $900 million of unconsolidated overhead expense by the government sector. This is to make no attempt to measure the revenues, costs, and taxes of Mexico's invisible oil industry—the club of Mexican firms whose principal function is to provide goods and services to Pemex.

This means that no fully understood, comprehensive general economic analysis of the government oil industry is possible, given the absence of financial data. What can be undertaken is the analysis of the operating and investment costs and income of Pemex, which supposedly are tracked routinely by Pemex itself but which are not made public.

Such analysis implies a knowledge of the principal published sources of Pemex industrial and financial data. Probably the best source on Pemex's production and financial data is a series published jointly by Pemex and the Ministry of Planning and Budgeting called *La industria petrolera en México*, whose numerous tables are understandable with little or no training in Spanish. Other sources for such data include Pemex's annual reports published on March 18, the statistical appendices to the President's state-of-the-nation address on September 1, and a number of in-house reports such as the report to Pemex's Administrative Council (not to be equated with a U.S. board of directors), issued bimonthly. Quite available and useful is the technical journal *Energéticos,* published monthly (during López Portillo's time) by the Technical Secretariat of the Commission on Energy Resources. The American Chamber of Commerce of Mexico periodically publishes a book-length update on the oil industry called *Energy Mexico,* which John Christman edits.

The Mexican press seldom publishes pieces of investigative journalism, but there are journalists, such as Raul Prieto, who specialize in reporting on the industrial and financial inefficiency of Mexico's oil industry. During

1981–1982, Prieto led a one-man attack on the management performance of Pemex during Díaz Serrano's tenure, putting much of it in his book, *Pemex muere* (Pemex Dies).[1]

The English-language press in Mexico, mainly aimed at the tourist market, occasionally reports on management issues in the oil industry. One article illustrative of Mexico City oil industry journalism appeared in *The News* of Mexico City, 22 November 1982. It discusses why the next government had good reason to create a new Energy Ministry, breaking up Pemex in the process. The reporter, Finance Editor Patricia Nelson, writes, "One of the reasons for breaking up Pemex would be to offset the strength of the union which has been able to command extraordinarily high wages throughout the industry, is a prime example of featherbedding and corruption, and many of its leaders hold the franchises on service stations."

A number of Mexican academicians write about oil policy matters, but only a very few examine industrial and management issues. Jaime Corredor published an exceptionally valuable collection of tables on the history of Mexico's oil industry, but his analysis was largely confined to noting growth rates of industrial output. Professor Jacinto Viqueira, at the Engineering College of the National University, carried out a study of efficiency in the energy sector, examining topics such as the implicit cost of flaring of natural gas.

In López Portillo's time, it was common for Mexican scholars to address U.S. academic, government, and business audiences at seminars and in scholarly reports.[2] A number of U.S. universities, notably Stanford under Clark Reynolds and the University of California San Diego at La Jolla under Wayne Cornelius, were hosts to these academic speakers from Mexico. It was not self-evident that these scholars had a corresponding constituency or audience in Mexico.

One historian, José Valero Silva from the National University (UNAM), was hired by Díaz Serrano to establish a historical research center in Pemex that included copies of all government and corporate archives abroad that touched on Mexico's oil industry during the period up to and including the expropriation of the oil industry in 1938. His research, however, was more in the area of corporate and government diplomacy and did not touch engineering or management issues in the narrow sense of the term. The program was terminated shortly after Díaz Serrano's departure from Pemex, leaving Valero Silva with a mountain of rare historical documents but no sponsor for his research.

---

[1]Edward J. Williams, *The Rebirth of the Mexican Petroleum Industry* (Lexington: 1979), pp. 20–21, names Herberto Castillo as the sharpest critic of government oil policy.

[2]A good example is Miguel S. Wionczek ["Mexico's Energy Policy—the Past and the Future," *Journal of Energy and Development* VII (2):199–212], who argues that the development of Mexican oil policy was a much more heated, diverse, and vigorous political exercise than "naive, foreign Mexico watchers" would assume from official documents or "scant local press reports."

There is at least one technical manual explaining the ins and outs of financial reporting at Pemex's level. The work is precisely focused on the complex issues of cost accounting in the Mexican petroleum industry. Translated, its title is *Manual for the Analysis and Interpretation of Financial Statements of the Petroleum Industry*. The book, published by the IMP in 1977, discusses intracompany cost-accounting issues at the plant and regional levels—practically down to the individual well. The source gives possibly the only published discussion of the principles that underlie—or at least should underlie—financial statements that Pemex prepares for its foreign bankers.

Unfortunately, given the consultant-client relationship between the IMP and Pemex, one should not expect to find—and indeed one does not find—any internal evidence in the book that Pemex actually carries out the prescribed procedures. Moreover, the book is written as if intended for chemical engineers who had never taken a course in business administration or accounting. Its authors therefore made no attempt to inquire into or offer historical cost data or analysis.

The book is a good deal prescriptive but, nonetheless, is a window into financial reporting and cost accounting at the middle-management level of the company. Because the text of the manual is informative about its subject as well as suggestive about how this subject is approached in Mexico, it seems worthwhile to review the main points that deal with the most difficult parts of Mexican financial reporting in the oil industry.

The Income Statement

The American income statement or profit and loss (P&L) statement is called the "statement of results" in Spanish. For being a para-state company, owned nominally by the Mexican people, Pemex is not supposed to smell too much like a normal, profit-motivated company. For this reason official documents, including those that deal with financial reporting, refer to the services rendered by Pemex and seldom speak in terms of outright profits and losses. An exception is the article in *The News,* which reported that in 1981 Pemex's gasoline refinery costs were 173 pesos/bbl while natural gas production costs were 72 centavos/cu m. The reported after-tax income on gasoline was 137 pesos/bbl and the net return on natural gas was 25 centavos/cu m. "At least part of the blame [for Pemex's low profits]," the author wrote, "can be placed on the so-called fictitious economy of unrealistically low retail prices of these commodities."

Often in Mexico one finds the concept of profit treated as merely the excess of revenues over expenses. The IMP manual refers to the income statement as "a dynamic financial document that details the results of the operations of a business during a specific period." The statement is made up of the four traditional groups: income, costs of goods sold, operating expense, and miscellaneous income and expenses. The IMP defines income as the value of the sale of goods and services that the firm produces. Ordinarily, this value is the market value; but in the case of operating units within Pemex, the value of an intermediate product is the value assigned from headquarters.

Cost of sales

This groups the cost of production plus the difference between initial and final inventories and the value of products purchased for resale. The difference between cost [*costo*] and expense [*gasto*] has long been a topic for polemical discussion.

Some professionals argue that cost is recoverable by the sale price, while an expense is not. Nevertheless, the excess of the sales price over cost is not all profit because part is required to pay for costs of operations. In IMP's study, cost will be the value of the diverse activities necessary to produce goods for sale, such as labor, primary material, electric energy, depreciation, and supervision. In contrast, expense will be the value of the multiple activities necessary for the functioning of departments or groups that distribute and sell the products, as well as those groups that offer services of a general character for the firm, such as administration, finance, personnel, supplies, and medical and legal services.

Operations expense

This group consists of the expenses of distribution and administration, such as salaries and services, advertising, maintenance of delivery vehicles and equipment, invoices, remittances, and other items related to sales. In addition, this category includes salaries, fees, telephone service, postage, depreciation of office equipment, accounting, collections, and other functions that form part of the administrative expenses.

Miscellaneous income and expenses

These are revenues or expenses arising from activities other than those associated with the principal activity of the firm. Such items occur irregularly and might include the sale of depreciated assets, interest charged or paid, and discounts on invoices or anticipated payments. The difference between income and the cost of sales is called gross profit, and this, less operations expense, is called operating profit. Miscellaneous income and expenses are then applied to operating profit to give the figure that serves as the basis for calculating taxes and profit-sharing among workers.

In Pemex, a consolidated P&L statement is prepared two times a year, June 30 and December 31 (Table 8.1).

In its annual reports, Pemex follows the practice of reporting only cash flow, in descending order of magnitude, mixing income-statement and balance-sheet figures freely. In its *Memoria de labores, 1981,* Pemex reported that it

Table 8.1
Consolidated income statement for Pemex operating unit

(Period Covered)	
Sales	$
Less:	
Cost of sales	$
Gross profit	
Less:	
Operating expenses	$
Operating profit	$
Plus:	
Misc. income	$
Less:	
Misc. expenses	$
Taxable income	$

Source: Instituto Mexicano del Petróleo (IMP).

"captured financial resources which, including gross financing, came to a total of 869,736 million pesos. The composition of the total is as follows: 471,773 of its own revenues, which represent 55%, and 397,963 million pesos of financing, which represents 45%."

As for "outflow," Pemex reported that during the same period costs and expenses "rose to 861,481 million pesos, made up in the following form [in millions of pesos]: federal tax, 238,193 (28%), investments, 230,733 (27%), debt retirement, 165,657 (19%), operational expenses, 134,217 (15%), interest payments 61,740 (7%) and other, non-associated expenses [*operaciones ajenas*]." This latter, somewhat mysterious category included items such as administrative loans to workers, the construction of worker housing units, down-payments on platform construction, value-added taxes paid, and "7,380 for expenses of operations and investments."

In its statistical yearbook, *Anuario estadistico,* published in October for internal use, Pemex follows the normal practice of presenting financial data in tabular form, showing amounts in billions of current pesos and the percent that each category of revenue—domestic sales, exports, and other—is of total revenues.

As for expenses, taxes, and profits, a slightly different practice is used. The sum of these three cateogries is taken as 100%, and each item's share is a percent of this figure. This procedure in effect understates the government's share in Pemex's operating income. In 1976, for example, domestic sales represented 53.3% of Pemex's total revenues, exports 10%, and other income 0.8%. In 1981 domestic sales were 24%, exports 75%, and other 1%. In 1976, total costs and expenses were 79%, leaving pretax profits of 21%, the government taking 20% of the total in taxes. Pretax profits were 10.1 billion pesos, from which the government took 9.7 billion in taxes, i.e., a 96% tax rate.

The technical manual, meanwhile, presents a traditional picture of the operational profits of a unit within Pemex (Table 8.2).

Table 8.2
Profits and losses for Pemex operating unit

(Month XX and Year)	*Month*	*YTD*
Value of the production	$	$
Less:		
Cost of production	$	$
Gross profit	$	$
Operating expense		
Packing, shipping and storage	$	$
Local administration	$	$
Operating income	$	$
Financial cost	$	$
Other products	$	$
Other expenses	$	$
Net income	$	$

Source: IMP

Drilling costs

Percentages of success can be determined by relating the total of productive and unproductive drilling to the total cost of drilling. The average cost per well can also be determined by dividing the cost of the activity by the number of wells drilled. The cost per meter drilled is determined by dividing the cost of drilling by the total depth drilled. By following the IMP guidelines, the general superintendents and zone and branch managers can use this information to facilitate technical and administrative control in the areas for which they are responsible.

The IMP manual points out that, because of the importance of this activity, drilling cost is handled as a specific asset account—alternatively, as a subaccount of "work in progress" and not as part of the cost of operation of the fields. As the manual intimates, the major use of this procedure becomes evident when transferring the value of productive drilling to its corresponding fixed asset account, namely "Wells." Further, when the value assigned to unproductive drilling is transferred to the "Reserve for Exploration and Depletion," the remaining category, "work in progress," would stand alone as an investment—since its ultimate placement is unknown until drilling is completed.

Extraction of crude oil and gas

Since this function also controls the transportation of the hydrocarbons extracted to the transformation centers, the separate costs of transportation and exploitation, which include extraction, gathering, and storage, need to be determined.

The information to formulate the statement of costs inherent in the activities are in the cost records compiled by the different accounting departments but containing excessive details for administrative officers. The IMP recommends a format for the production statement that eliminates unnecessary detail (Table 8.3).

Table 8.3
How Pemex calculates cost of production

(Period XX)	*Month*	*YTD*
Initial inventory		
Plus:		
Direct labor	$	$
Materials	$	$
Fuels, grease, and oils	$	$
Repairs	$	$
Indirect labor	$	$
General charges (depreciation, amortization, additions to reserve accounts, pensions)	$	$
Other costs	$	$
Total	$	$
Less:		
Final inventory	$	$
Production cost	$	$
Plus:		
Cost of exploration	$	$
Cost of oil-field operations	$	$

Source: IMP

Each component of industry reports on its cost of production. The statement of production costs therefore includes the elements necessary to control the input used, or the process of manufacturing goods and providing services. The statement of production costs is the basis to determine the unitary cost of the products and is used to establish operation efficiency indices, integrated with the cost statement and the income statement. The main financial information is used by the general manager. The statement gives the value of the elements used in the production, transformation, or manufacture of goods in a period that, depending on the manufacturer, may be intermediate or final products whether they are to be sold or are for producing another product in the same business.

This statement is composed of the traditional concepts of production: raw material, labor, and indirect charges (overhead). The first two are the easiest to identify in the transformation processes and consequently are the easiest to quantify for each product. The third group receives the name indirect because it is impossible to quantify the participation in these charges in each item. For example, indirect charges include supervisor salaries, electricity, fuels, lubricants, and paperwork. They also include other charges that do not imply expenses, for example depreciation and amortization, but which are indispensable to production and are also put in the category of production costs.

At the end of the period covered by a statement, some items are still in production and will not be completed until the following period. The current value of these items must form an integral part of the financial statement in progress. The production in progress at the end of the former period is managed as initial inventory of the manufacturing in progress, and the value of the production at the end of the period is treated as final inventory.

The cost of production at Pemex is made up of the cost of goods sold. To integrate these costs, one refers to the information partially contained in the cost records prepared by the primary production areas (Northern and Southern zones and the Poza Rica and Northeast Border Districts) and the industrial production centers (refineries and petrochemical complexes). This information is "partially" contained because, to consolidate the operations of Pemex, company accountants increase or decrease the value in the cost records in the process of quantifying the cost of depreciations, amortizations, and the increments to several reserve accounts that appear on the company's balance sheet.

Unit costs

The corporate accounting department in Pemex annually obtains the unit costs for each of the production centers. Such information is found in two volumes: *a*) unit costs of crude oil and natural gas and *b*) unit costs of refined products and petrochemicals. These cost records contain information about the raw materials that are used in each plant, as well as the final and intermediate products obtained. Data on raw materials are given only in volumes, not in values. The records also cover the cost of direct and indirect labor. For each production district and petrochemical refinery plant, they cover the value of the fuels, grease, oil, and local services, as well as storage, packaging, shipment costs, and expenses for general services, such as personnel, medical services, accounting, and data processing.

A Mexican petroleum engineer interviewed in 1982 stated that Pemex generally has not formally maintained a staff section dedicated to production cost analysis. The comment was made that when President Miguel de la Madrid was Vice President of Finance at Pemex, some effort was made to generate financial statistics measuring industrial efficiency. But upon his reassignment, the program was dropped.

Whatever the facts, the Mexican government in general and Pemex in particular abhor publishing unit cost statistics but revel in production data—"an obsession," the engineer said, "bordering on demagoguery." The only unit cost data published are prices to a Mr. Al Publico, who turns out to be the general public. Even then, revenues received by Pemex on account of the sale of a liter of gasoline, ex-dealer commission and ex-taxes, are not given.

From an accounting point of view, the cost records show the cost of operations (factory cost) to which the value of the raw material is added to obtain the cost of production. They also show costs of packaging, storage, shipment, and local administrative expenses.

For Pemex as a whole, the packaging, storage, shipment, and administrative expenses of each production center are all part of the cost of production. The activities in the working centers are for this reason registered in the records of the field operations, refineries, and petrochemical plants. Separately, each producing district and each processing center maintains records on its costs of production and operational expenses. Their financial statement distinguishes between concepts that aggregate the cost of production and concepts that are not directly related to production activity, such as distribution and administrative costs. The latter two should form part of the local operation costs and therefore are included in the local profit and loss statement.

The cost records include, among other information, the cost of operations and the volumes of raw materials processed (crude oil, gas, and intermediate processed products). To determine the total cost of production, these costs plus the value of the initial and final inventories of the period are all that is needed. To carry this out, reference is made to the certified accounting reports and to the unit prices of the different products established yearly by the accounting department of Pemex.

Because Pemex is involved in diverse activities ranging from exploration to distribution of its products, the IMP recommends that cost statements in each work center differentiate among its various activities. This implies that each processing plant formulates a separate cost statement for refining and petrochemical activities. Each producing zone also reports separately on the costs of its several functions, which include drilling, production, refining, and petrochemicals. (Exploration costs are covered under field production costs.)

Drilling

This cost statement gives information relative to the cost of drilling in the current period. This is obtained through gathering elements that are entered in the cost records, which include categories for exploration, drilling, production, gathering, and transportation. These need to be regrouped according to the activity that is being analyzed in order to integrate the relative costs (Tables 8.4 and 8.5). The IMP manual suggests that information given in the respective cost statements be complemented with the information of the previous month in order to catch discrepancies between the periods so solutions can be found at an early stage. The recommended format of the cost statement includes a column that shows the year-to-date (YTD) values of expenses.

Table 8.4
How Pemex calculates cost of completed drilling

(Period XX)	*Month*	*YTD*
Drilling in initial stage		
Exploration	$	$
Development	$	$
Total	$	$
Plus:		
Labor	$	$
Pipelines	$	$
Materials	$	$
Fuels, grease, and lubricants	$	$
Repairs	$	$
Indirect labor	$	$
Depreciation, amortization	$	$
Other	$	$
Contract drilling	$	$
Total	$	$
Less:		
Final drilling	$	$
Exploration	$	$
Development	$	$
Cost of completed drilling	$	$

Source: IMP

Cost of exploration

The IMP comments that traditionally the cost of exploration has been considered an expense or a debit to the reserve for exploration and depletion of the fields. Nevertheless, even when the amount of the expenses for exploration is not of the magnitude of the operation of the fields, refineries, petrochemicals, etc., "it is indispensable to formulate an analytical framework that helps the

Table 8.5
How Pemex calculates drilling in progress

(Period XX)	*Month*	*YTD*
Cost of completed drilling	$	$
Less:		
Productive drilling	$	$
Exploration	$	$
Development	$	$
Total	$	$
Less:		
Unproductive drilling	$	$
Exploration	$	$
Development	$	$
Cost of drilling in progress	$	$

Source: IMP

superintendents of exploration in the Northern and Southern Centers and the districts of Poza Rica and the Northeast Frontier understand the cost information supplied."

The cost records of the exploration centers are analyzed in the same way as other activities, namely, direct and indirect labor, fuels, grease, oil, etc. The details of cost are functional instead of conceptual but retain enough detail so that different items of expense are not summarized in one or two categories. The conceptual or functional analysis of this sector, in addition to informing the superintendents of exploration and the branch managers of what goes on in their corresponding jurisdictions, facilitates the planning process.

Similar to the cost of extraction statement, except for the final and initial inventories, the cost of exploration statement is prepared for each district (Table 8.6).

Refined products and petrochemicals

The industrial production activities of Pemex are accountable and, with the exception of the peculiar terminology used by the industry, are similar to industrial firms of the private sector. Monthly and year-to-date records on costs are maintained by individual refineries and petrochemical plants. Since the refineries also manufacture petrochemical products, a separate cost of production statement is made. This information is also in the cost records.

Selling expenses

This is a detailed statement of the costs of selling, which includes depreciation, amortization, taxes, increments to reserves for missing or obsolete

Table 8.6
How a Pemex district reports cost of exploration

(Period XX)	*Month*	*YTD*
Geophysical		
Gravimetric	$	$
Magnetometric	$	$
Seismographic	$	$
Central units	$	$
Field crews	$	$
Other	$	$
Total	$	$
Geology		
Subsoil	$	$
Surface	$	$
Paleosedimentation	$	$
Central units	$	$
Other	$	$
Total	$	$
General and administration expenses	$	$
Total cost of exploration	$	$

Source: IMP

inventories, indemnities, and pension plans for each of the subsidiaries of sales, such as terminals, shipment centers, and agencies. For the administrative authorities to become informed of subsidiary costs, the sales accounting office is advised to make monthly statements for the operation of each of them with the corresponding year-to-date accumulation.

The sales agency tabulates the consolidated information of its dependents (agencies, terminals, embarking terminals, plants, and central offices that it controls), presented as expenses. Individual statements of each of the sales subsidiaries should be made for the corresponding company officers where the expenses of their jurisdiction are detailed in accordance with the budgets and the distribution of indirect and other charges that correspond to each subsidiary, such as labor, materials, medical services, freight charges, taxes, depreciation and amortization, additions to reserves, shortages, obsolete inventories, indemnifications, and retirement. The cost of selling statement is prepared by the sales accounting office.

Maritime cost

This statement is a list of subconcepts used in operations that produce services instead of products. The total cost of its operations is calculated and the amount is distributed among the branches that used the service. Thus, the balance at the end of each year is zero.

Nevertheless, a monthly cost report of maritime operations is prepared to inform the branch manager and superintendent about the cost of the maritime activities. Because of the extensiveness of the geographic area in which the activities take place, the IMP technical manual suggests that the report be done as in the sales branch, showing cost by concept, such as marine terminals, superintendencies, marine units, and national and international transportation.

To evaluate the efficiency of each operation a statement is prepared for each trip. In addition, the distribution of the maritime cost by Pemex department is prepared (Table 8.7).

Table 8.7
How Pemex charges itself for maritime costs

(Period XX)	*Month*	*YTD*
Pemex department		
Production	$	$
Refining	$	$
Petrochemicals	$	$
Sales	$	$
Domestic	$	$
Export	$	$
Total	$	$

Source: IMP

Construction

The IMP's point to petroleum engineers is that, although projects and construction do not constitute a specific activity of the petroleum industry, cost financial control is important because of expansion in recent years. What is equivalent to a statement of production, namely the cost of construction, covers the three largest elements of cost: labor, materials, and general expenses of construction. The value of the statement is that its analysis, when combined with the budget for each project in its different stages of construction, allows control of construction cost conceptually and by project. The accounting by month of construction and projects follows formats given earlier (Table 8.8).

Investment criteria

This review of cost accounting precepts dramatizes, first, the educational role of the IMP in Mexico's petroleum industry. The Institute's work in this area is through publications as well as through an extensive seminar program for both professionals and union workers. This review also shows how IMP and Pemex

Table 8.8
How Pemex accounts for project construction costs

(Period XX)	*Month*	*YTD*
Works under construction (initial)	$	$
Materials	$	$
Direct labor	$	$
Diverse charges		
Fuels, grease, and oils	$	$
Electric energy	$	$
Equipment leasing	$	$
Shipping and hauling	$	$
Transportation	$	$
Interest payments	$	$
Depreciation	$	$
Amortization	$	$
Additions to reserves for pensions and indemnities; losses and obsolete equipment	$	$
Total	$	$
Less:		
Works under construction (final)	$	$
Cost of works completed	$	$

Source: IMP

managers perceive the intellectual problem of cost accountability in the operational side of their industry.

Unfortunately, no clue is given to the criteria—above all, profitability—for investments in production, transportation, refining, or petrochemicals. The IMP manual does give examples of the financial statement of a hypothetical Pemex refinery with an operational profitability of less than 2%. This figure, however, cannot be taken as a measure of true profitability, owing to the complex system of subsidies that distorts costs and revenues.

Besides, even where true profits from annual operations are known, the question of the return on capital and other investment-analysis issues are left unaddressed. The reason is no doubt because only a handful of financial analysts, many of whom are likely not to be in Pemex, are given the problem of deciding if, when, where, why—and with how much outlay, risk, and profit—a new refinery is to be built. Once the plant is built, the precepts outlined by the IMP staff take over. But before that the petroleum engineer in Poza Rica is likely to be as much in the dark as anyone on the U.S. side of the border with regard to investment criteria employed at the Finance Ministry, the President's office, and at Pemex headquarters concerning next year's drilling, production, or refinery investments.

Pemex Financial Management, 1976–1981

This financial section deals with Pemex as an agency of the Mexican government (analogous to the TVA in the United States). As an executive agency, Pemex carries out various functions (and bears expenses) not normally associated with a privately owned, integrated oil company (Figure 1.1). Therefore, the agency's income (Table 8.9) and expense in per-barrel terms (Table 8.10) should be viewed in this light.[3]

From 1976–1981, Pemex's income rose tenfold while current costs and expenses, including financial, rose only by a factor of 6. Gross profits increased 24½ times but taxes increased over 25 times, permitting operating income to increase only by a factor of 3. Net profit at the end of the period had increased by only 43% from the level of 1976, but employee profit-sharing had gone up 11 times.

The bible of any analysis of Pemex and Mexican oil finances shows *audited* peso income and expenses for the 6-year period under review (Table 8.9). No audited dollar figures for Pemex exist because the company is peso-based, meaning that its account is credited in pesos at the current exchange rate for its export sales revenues, which are paid in dollars. Similarly, Pemex buys dollars at the current exchange rate from the Bank of Mexico for its imports and dollar-denominated contracts in Mexico. As a peso-based company, Pemex has experienced both gains and losses in its foreign currency transactions.

Pemex's exports during the 1976–1981 period increased in value at an annual compound rate of nearly 120%, while the cost of operations increased by less than 43%. The resulting annual increase in gross profits in excess of nearly 90% was matched by the growth of federal taxes. The table also shows the indirect result of Pemex's ability to leverage Mexico's crude reserves in financial markets: net payments for financial services rose by over 86%/year.

Pemex's management succeeded in *reducing* the ratio of current expenses to net sales at an average rate of more than 10%/year, with the result that the ratio of pretax income to net sales rose at an annual rate of nearly 20%. Federal taxes as a share of pretax income stayed at an average of 98%. Employee profit-sharing rose at an annual rate of almost 62%. Net profit, meanwhile, was never more than 1% of net sales after 1978.

Pemex increased its crude oil production at an annual rate of almost 25% while increasing its sales revenues per barrel by 28%. Costs, however, were

---

[3]Pemex's income statement can be reorganized to approximate that of a private company (Table B.1). Bankers do not seem to object to Pemex's practice of lumping nearly all of its expenses into "cost of sales." Pemex's foreign debt in 1983 of $25 billion, greater than that of most developing countries, reflected banker confidence in Pemex's credit worthiness, regardless of dubious financial reporting practices. With the assumption of credit-risk responsibility for Pemex by the U.S. government, financial reporting requirements were relaxed to the point that in 1983 Pemex was able to obtain $100 million in credits via the Exim Bank without having to submit 1982 financials.

Table 8.9
Pemex's 6-year financial performance in current pesos*

	1976	*1977*	*1978*	*1979*	*1980*	*1981*	*1976–1981 compound growth rate†*
Revenues							
Exports	6,834.9	23,723.3	41,898.2	92,876.2	239,136.0	349,284	119.6
Domestic sales	39,634.2	53,005.8	58,696.4	73,177.0	96,324.8	111,640	23.0
Net sales	46,469.1	76,729.1	100,594.6	166,053.2	335,460.8	460,924	58.2
Other income	800.0	935.6	1,234.9	1,271.8	4,766.9	4,757	42.8
Total	47,269.1	77,664.7	101,829.5	167,325.0	340,227.7	465,681	58.0
Cost and expenses							
Cost of sales	30,850	39,707.3	52,288.0	74,841.9	133,268.1	151,265	37.4
Dist. costs	4,657	7,895.8	9,995.1	15,762.9	17,583.4	29,937	45.1
Financial costs	1,486	3,071.1	4,754.7	13,685.1	15,227.2	33,174	86.1
Other expenses‡	249	6,801.2	3,925.6	(554.9)	4,522.5	5,988	88.9
Total	37,242	57,475.4	70,963.4	103,735.0	170,601.1	220,364	42.7
Pretax income	10,027	20,189.3	30,866.1	63,590.0	169,626.5	245,317	89.6
Federal taxes	9,661	19,764.5	30,258.2	62,886.5	168,675.4	244,179	90.8
Operating income	366	424.8	607.9	703.5	951.1	1,138	25.5
Employee profit sharing	63.6	90.1	220.5	373.0	553.4	705	61.8
Net profit	302.4	334.7	387.4	330.5	397.7	433	7.4

*Millions of audited pesos
†Shown as annual growth rate (%)
‡Includes exchange-rate (profits) and losses.
Source: "Pemex as a Financial Entity," *Energy Mexico, 1981* (Mexico City: Amcham, 1981) for Pemex financial data and audited income statements.

Table 8.10
Pemex's per-barrel revenues, operating costs and profits, 1976–1981 (current pesos)

	1976	*1977*	*1978*	*1979*	*1980*	*1981*	*6-year* average	*Compound growth rate*
Crude oil and condensates (Million bbl)	292.3	358.1	442.6	536.9	706.7	844.2	530.10	23.6
Sales rev./bbl*	158.98	214.27	227.28	309.28	474.69	545.99	321.74	28.0
Cost/bbl**	127.41	160.50	160.33	193.21	241.41	261.03	190.65	15.4
Pretax income/bbl†	34.30	56.38	69.74	118.44	240.03	290.59	134.91	53.3
Federal tax income/bbl	33.05	55.19	68.36	117.13	238.68	289.24	133.61	54.3
Pemex operating income/bbl	1.25	1.19	1.37	1.31	1.35	1.35	1.30	1.6
Net profit/bbl	1.03	.93	.88	.62	.56	.51	.76	−13.1
Net increase in reserve/bbl‡	12.82	17.90	10.42	8.05	49.51	10.79	18.25	− 3.4
Net increase in national patrimony/bbl§	46.90	74.02	79.66	125.80	288.75	300.54	152.61	45.0

*Revenue includes sales of crude oil, refined products, petrochemicals and natural gas.
†Corresponds to gross profit before federal taxes in Table 8.9, which includes the effect of other income.
‡For unproductive exploration and depletion.
§The sum of federal tax income/bbl, net profit/bbl, and net increase in reserve/bbl.
**The agency's loaded cost per barrel, including cost of goods and selling, administration, and financial costs.
Source: Table 8.9 and its source.

held to a growth rate of only 15%, permitting pretax income (and taxes) to rise in excess of 53%/year. Since Pemex is expected to generate profits for the nation and not for itself as a management entity, it is not surprising that as production rose net profit per barrel should have fallen, as it did, at an annual rate of 13% (Table 8.10).

While 1980 marked a milestone of vigorous financial performance for Pemex, 1981 performance faltered under the weight of a heavy millstone—the oil glut of 1981. This is not to say that growth stopped; rather, it slowed dramatically. In 1976, domestic sales accounted for 83.5% of net sales, with exports amounting to 14.5%. By 1981, this relationship had nearly reversed, with exports becoming the overwhelmingly dominant generator of income, contributing 77% of Pemex net sales. Thus, Pemex's and Mexico's petroleum-related earnings are highly sensitive to external market forces.

Export revenues in 1980 follow from the 1976–1980 compound annual growth rate of 143.2%. Had the growth rate continued through 1981, export earnings would have been 224 billion pesos more than was actually the case. Export earnings for 1981 were, therefore, 38.5% less than their projected value using the 1976–1980 base line (Table 8.11).

However, total income in 1981 was only 16.4% less than projected using the 1976–1980 growth rate. Using the 1976–1980 base line in this case masks the sharp drop in the growth rate experienced from 1980–1981 due to the oil glut. The increase in total income from 1979–1980 was 103.3%. The same figure for 1980–1981 shows only a 36.9% net increase, or a 64.3% drop in the growth rate of total income. This corresponds very closely to the 68.6% drop in the growth rate of export income from 1980–1981. Total income suffered slightly less than export income because domestic sales, while not as strong in 1981 as in 1980, slowed less dramatically than did export sales.

Pretax income in 1981 was 28.7% less than its projected value, while total income dropped 16.4%. (Percentage change between 1980 and 1981 growth rates is −73% for gross profit and −64.3% for total income.) The pretax income figure fell further than total income because the growth rate in costs and expenses for 1981 fell less from their projected level than did any other category in the income statements. Costs and expenses were only 11.7% off their projected total, thereby maintaining a high level relative to the drop in total income. They therefore took a relatively bigger bite out of total income in 1981 than they did during the 1976–1980 period (Table 8.11).

Pemex and Government—Financial Embrace

Before reviewing Pemex's revenues, costs, and profits for 1976–1981, the company as a mangagement entity should be placed in its proper context. The line between company mangement and government management for NOCs (national oil companies) is, in the public's eye, hazy and sometimes nonexistent.

Table 8.11
Pemex's 1981 financial performance

Item	*1976–1980 Compounded annual growth rate, %*	*1980**	*1981 Projected value per 1976–1980 base line**	*Reported 1981**	*Projected/reported var., %*
Exports	143.2	239,136	581,578.8	357,537.8	−38.5
Total income	63.8	340,227.7	557,292.9	465,681	−16.4
Total costs and expenses	46.3	170,601.1	249,589.4	220,364	−11.7
Pretax income	102.8	169,626.5	344,002.5	245,317	−28.7
Taxes	104.4	168,675.4	344,722.5	244,179	−29.2
Net profit	7.1	397.7	425.9	433	+ 1.7

*Million current pesos.
Source: Pemex, *Memoria de labores,* 1981, p. 121

Drawing this line to measure management performance is at times difficult. For example, the NOC may have actual costs that, for accounting purposes, are absorbed by other government agencies and budgets. At the same time, the NOC may be asked to absorb costs that a business organization would not necessarily incur. The NOC may be required to maintain a certain payroll level, in dollars and personnel, that could only be justified in the light of government welfare and employment policies. The NOC's net profits, after taxes and employee profit-sharing, may be much lower than other indicators of management performance might indicate. Such poor net profit performance appears, as an accounting phenomenon, because of the high effective tax rates imposed by the NOC government.

The NOC is like any state-owned company in having a public policy role. The social and strategic role of the public enterprise, as understood by the president and his economic cabinet, is a related but quite distinct issue from that of administrative and asset-management performance at the company level.

The NOC should therefore be evaluated as if it were a privately owned company for those areas in which its managers have operating responsibility. To measure the efficiency and profitability of management performance in those areas is not to measure the results of the overall economic policies and programs. But to the extent that the success of national policies depends on the success, financial and industrial, of the NOC, an adequate analysis of NOC performance becomes the point of departure for any appraisal of economic development strategy.

Pemex's foundation as a public institution rests on an article in Mexico's Constitution of 1917. Pemex, whose origins go back to the expropriations of 1938, is the State's oil exploration, production, refining, and wholesale distribution company. The State, not Pemex, is the owner of Mexico's oil reserves, so in this respect Pemex differs from the U.S. integrated oil companies that own or lease a good portion of their oil deposits.

Finally, there can be seen a basis for a preliminary discussion of Mexico's national participation per barrel—a Venezuelan oil-industry term. The term refers to an estimate of the total net income, measured in units of domestic purchasing power, that the nation receives for each barrel of crude oil produced. In Venezeula, income taxes paid by foreign oil industry service companies for technical assistance are added to sales revenues, again on a per-barrel basis.

In the absence of corresponding data, the Mexican figure is calculated as the sum of federal taxes, net profit, and the increase in Pemex's depletion reserve. The resulting calculations, which ignore the value of economic multipliers, therefore understate the total value of Pemex's operations to the nation. Table 8.10 shows that Pemex's contribution to national income per barrel increased more than six times to nearly 300 pesos/bbl from less than 50 pesos/bbl.

Table 8.12
Balance in Pemex's current account, 1976–1981

(Millions of dollars)

	1976	*1977*	*1978*	*1979*	*1980*	*1981*	*Compound growth rate*
Revenues							
Exports	436.1	1,018.8	1,837.2	3,986.5	10,401.9	14,585.0	101.8
Cost and expenses							
Petroleum imports	229.9	208.2	307.6	540.2	765.9	682.6	24.3
Other imports	524.6	460.3	979.0	957.4	1,129.2†	1,534.0	23.9
Interest payments	168.3	168.5	324.5	580.2	642.5†	1,353.5	51.7
Imports of services‡	22.6	25.9	30.0	112.8	NA‡	NA‡	—
Total	945.4	862.9	1,641.1	2,190.6	2,537.6	3,570.1	30.4
Balance	(509.3)	155.9	196.1	1,795.9	7,864.3	11,014.9	189.9*

†Estimated.
‡Not available after 1979.
Source: Pemex; J. Corredor, "Petroleo en Mexico."

Pemex's dollar income and expenses in current dollars

While Pemex's peso-denominated income statements show substantial increases in gross revenues and taxes, they understate the importance of Pemex's commercial operations for the economy in general and the government in particular. Pemex's operations created much greater dollar purchasing power for the country than peso-reported income statements would suggest. Sales earnings were increasingly in dollars—to 72% in 1980 from 14% in 1976.

While annual figures for the aggregate demand by Pemex for dollars for both operations and capital expenses, including leasing, are not available, some approximation of Pemex's current account is possible. Pemex succeeded in increasing its dollar revenues at an annual rate of 121% while limiting the growth of its current account dollar requirements to less than 30%. The result is that Pemex's current account balance, which began with a deficit in 1976 and increased to a surplus in 1977, rose at an annual rate of 650%. A half-billion-dollar deficit in 1976 was in excess of a $7 billion surplus in 1980 (Table 8.12).

Current income and expenses adjusting for inflation

Inflation must be accounted for in evaluating Pemex's financial statistics. Since the peso is Mexico's national currency, dollar revenues are useful only for imports; yet the purchasing power of the dollar generally has been depreciating, owing to inflation in the U.S. and other industrialized countries. At the same time, the peso's purchasing power in Mexico has been depreciating at an even faster rate. Owing to a nonparity exchange rate policy, however, the peso has been appreciating against the dollar in relative terms by the factor by which the peso has become increasingly overvalued against the dollar.

Pemex benefited from this situation by being able to pay for its imports with more cheaply acquired dollars, and the nation benefited by being able to credit Pemex's account in pesos, keeping the dollars for itself. This relative advantage for Mexican importers was the relative disadvantage of Mexican exporters for whom every dollar of export sales in 1980 produced only 22.95 pesos of income instead of 33.79, which as an exchange rate would have approximated the 1972 parity of purchasing power between the two currencies.

The result is that Pemex's real growth rate of export sales was 102%, not 121%, and costs increased 18% vs 29%. Debt service in Pemex's capital account consumed some of the dollar surplus indicated in the current account balance, but in 1980 more than $7 billion remained after all current account dollar expenses and capital-account interest payments had been paid (Table 8.13).

Expressed as the real value of peso-denominated revenues, domestic sales and operating income, for example, were virtually level in real terms throughout the period despite the 25–27% nominal growth shown. Similarly, cost of sales rose only 16% each year in real terms, not 44% as shown earlier, and employee

Table 8.13
Pemex's foreign purchasing power skyrockets in 1979 and 1980*

	(Millions of 1981 $)						
	1976	*1977*	*1978*	*1979*	*1980*	*1981*	*Compound growth rate*
Price deflators for imports of goods and services†	0.377	0.586	0.656	0.748	0.858	1.00	—
Deflated Pemex dollar revenues	1,156.8	1,738.6	2,800.6	5,329.5	12,121.8	14,585.0	66.0
Deflated Pemex dollar cost‡	2,507.7	1,472.5	2,501.7	2,928.6	2,957.2	3,570.1	7.3
Balance in 1981 purchasing power	(1,350.9)	266.0	298.9	2,400.9	9,164.6	11,014.9	153.7*
Pemex repurchase of dollars for capitalized interest payments§	201.5	148.4	213.1	365.9	703.1	1,103.1	40.5
Balance¶	(1,552.4)	117.6	85.8	2,035.0	8,461.5	9,913.6	203.0*

*1977–1981.
†Derived from Diemex-Wharton figures (1970 = 1.0)
‡Dollar costs in current account
§Current pesos converted to dollars at IMF annual average exchange rate and then inflated to 1981 values.
¶Available for Pemex's capital, goods, imports, and government general revenue purposes; 1980 and 1981 balances are net of current account services imports.
Source: Pemex, *Memorias de labores;* Dictamen y estados financieros, 1981; Diemex-Wharton.

profit-sharing rose 38% annually, not 72%. Net profit, meanwhile, actually fell 14%/year in real terms.

Pemex's real income in 1980 was 197% more than that of 1976, while real costs and expenses in 1980 were 89% more than in 1976.

Expressed in 1980 constant dollars, revenues per barrel rose to $20.65 from $16.76, while the cost per barrel dropped to $10.50 from $13.43. Pemex pretax income rose 30% to $10.44/bbl from $3.62/bbl. Federal tax income per barrel rose to $10.38 from $3.48, while net operating income per barrel predictably fell to 6¢ from 13¢. The net increase in national patrimony rose 26% annually to $12.55/bbl from $4.94/bbl, yielding a period-end growth of 154% from the level of 1976 (Table 8.14).

The figures are approximate and do not account for the fact that they would be lower if Mexico's natural gas production during the period were converted into equivalent barrels of crude oil. Crude oil equivalent production (not calculated) divided into actual crude sales revenues would considerably lower all figures. There were, throughout the period, large volumes of natural gas that were flared, generating no income. To some extent this shortcoming is compensated by including revenues from natural gas sales, domestic and export, in the total figure used to derive per-barrel revenues, costs, and profits.

This approach to the analysis of the management performance of an NOC permits conclusions in three areas: *a*) net profit per barrel calculations, *b*) internal management accomplishments, and *c*) government management policies.

Net profit per barrel calculations

There are three ways to calculate net profit per barrel.

a) NOC contribution to national income per barrel is a figure that represents on an annual basis the NOC's total contribution to the nation's income and capital accounts, measured in units of domestic purchasing power. To make real-value calculations, this figure is adjusted for the inflation experienced by the national economy. Pemex's contribution to national income follows this method of calculation.

b) NOC dollar cash flow per barrel is a figure that represents on an annual basis the NOC's total dollar revenues less its total dollar outlays (current and capital) divided by total crude production. The relative importance of this figure will vary with the strength of the dollar against other currencies, and the figure can be adjusted for OECD inflation to permit constant-dollar calculations. In the absence of data regarding Pemex's total dollar outlays, the profit or loss in dollar cash flow per barrel cannot be calculated at this time.

c) National participation per barrel is a measure of the total revenues, in units of domestic purchasing power, that accrue to the state in a given year as a result, direct or indirect, of NOC operations. This figure is calculated as the sum

Table 8.14
Pemex's adjusted per barrel revenues, operating costs and profits

	(1981 dollars)							
	1976	*1977*	*1978*	*1979*	*1980*	*1981*	*6-year average*	*Compound growth rate*
Crude oil and condensates (Million bbl)	292.3	358.1	442.6	536.9	706.7	844.2	530.1	23.63
Sales rev./bbl*	30.75	21.72	19.56	22.12	26.29	22.28	23.77	− 6.24
Cost/bbl†	24.63	16.27	13.80	13.81	13.30	10.65	15.41	−15.44
Pemex gross profit/bbl‡	6.63	5.72	6.00	8.46	13.24	11.86	8.65	12.33
Federal tax income/bbl	6.39	5.59	5.88	8.38	13.17	11.80	8.54	13.05
Pemex operating income/bbl	0.24	0.13	0.12	0.08	0.07	0.06	0.12	−24.21
Net profit/bbl	0.21	0.09	0.08	0.05	0.03	0.02	0.07	−37.52
Net increase in reserve/bbl§	2.48	1.81	0.90	0.57	2.74	0.44	1.49	−29.24
Net increase in national patrimony/bbl¶	9.08	7.49	6.86	9.00	15.94	12.26	10.11	6.19

*Revenue includes sales of crude oil, refined products, petrochemicals and natural gas.

†Corresponding to cost and expenses total line of Table 8.9.

‡Corresponding to gross profit before federal taxes line of Table 8.9, which includes the effect of other income.

§For unproductive exploration and depletion.

¶The sum of federal tax income per bbl, net profit per bbl, and net increase in reserve per bbl.

Source: Table 8.10; and Diemex-Wharton derived GDP deflators.

of the NOC's contribution to national income plus the value of economic multipliers such as income taxes paid by companies whose principal revenues derive from NOC sales or service contracts. In Mexico's case, such calculations have not yet been carried out by the national treasury officials, who have sole access to aggregate corporate income tax data.[4]

Standing on tiptoes to get a glimpse of Pemex's financial condition will be much less of a strain if the de la Madrid government continues in the trend set by its publication of Pemex's financial statistics for 1982. The promising subtitle of the documents detailing Pemex's income, expenses, and debt read "[A] Decentralized, Public Agency of the Federal Government." The bad news is that to emphasize Pemex's ties to the federal government is to head the prow of analysis toward largely unchartered seas.

---

[4]The exercise of restating Pemex's income statement in constant dollars of 1980 was carried out in the author's article, "The Pemex Boom—Adjusted for Inflation," *Energy Detente,* volume II (17), December 21, 1981.

9

Political management of the petroleum sector

Mexican Spanish has a large number of deceptive cognates. Words like *constitución, presidente, elección, congreso, justicia, suprema corte, cámaras de comercio, leyes, voto,* and *partido* connote a social reality that does not at all correspond to that reality which, for example in the U.S., would correspond to the English cognates of these words. A "congress" that has never refused to pass a presidential legislative initiative cannot be called the equivalent of the U.S. Congress. Similarly, a president who has never vetoed a congressional bill is not a president in the American political sense of the term. A political "party" that never loses an election is not a political party in the American sense, and a "supreme court" that never overturns presidential decrees or congressional or state laws or initiatives is not a supreme court in the American sense.

In Pemex there is a "labor union" called the STPRM that has the curious record, by the standards of American labor history, of never having gone out on strike since the expropriation of the oil industry in 1938. There is also a "director general" who, were Pemex established by law as a commercial entity governed by the Ley de Sociedades Mercantiles, should be the same as an American company president. But because Pemex is established as an agency of the Mexican government, no principles of organization or accountability automatically apply. As the removal of Díaz Serrano in June of 1981 illustrates, being director general does not necessarily carry with it even the authority that would be given to a third-level manager in a U.S. oil company.

These observations are not meant to suggest that the political and management systems of Mexico are better or worse than those in the U.S., nor that Mexico's political vocabulary is or is not suited to prevailing institutions and practices. The point is that Mexican terms and practices are—almost—*sui generis,* and any sense of familiarity brought about by the existence of cognates to Spanish terms is likely to be mistaken.

This in turn implies something about a non-Mexican's reading of the history of Mexico's petroleum industry. If Mexico's Constitution of 1917 is not at all a constitution in the American sense of the term, that means that the authority of the famous Article 27 giving the state subsoil rights should be questioned. If Mexican supreme courts only do the bidding of Mexican presidents, then the various findings of Mexican courts need to be questioned. If, similarly, Mexican labor unions only do as they are told, then the various demands of the unions in the months prior to the March 1938 expropriation (timed to be eclipsed in the media by one of Hitler's moves in Europe) need to be reexamined.

The Generic Character of U.S.-Mexican Relations

Speaking about the risk factor in loans to Pemex, a U.S. banker commented (in December 1982) that the problem of risk was not one that his bank intended to analyze closely given that, in the bank's view, "the last guarantor of a loan to Pemex was 'the full faith and credit' of the U.S. government." He didn't say the *Mexican* government. His comment suggests that, from a financial viewpoint, the Mexican economy is already annexed to the U.S. In other words, it was inconceivable that the U.S. would let a matter of debt endanger the political and military stability of Mexico and by implication affect the stability of the U.S.-Mexican border.

One might conclude, therefore, that Mexico's relationship to the U.S. is analogous to that existing between South Korea and Japan or, more distantly, between Poland and the Soviet Union. Where a nation borders on a potential predator nation, a common strategy for self-defense is ideological and institutional camouflage. The lesser power adopts the coloration, like a chameleon, of the local environment at that season of the year.

In Mexico's experience, with half its former territory in U.S. hands and with, during the period of Porfirio Díaz, much of its private industry in foreign hands, there were good reasons to stay beyond Uncle Sam's grasp. Mexico's solution to this central problem of national security was to achieve in constitutional and ideological terms such an imitation of the U.S. that Uncle—flattered to see his reflection in the Mexican political mirror—would leave Mexican territory and political and economic independence alone.

The U.S.-Mexican War settlement was supposed to be an answer to the questions: How shall further U.S.-Mexican relations be governed? and How shall Mexico be governed in the face of a permanent power imbalance between the two countries? American and British diplomats concluded that Mexico, because of its cultural, ethnic, and language differences, could never explicitly be absorbed into the U.S.; therefore, it should be left to be governed by Mexican upper classes with the implicit support of the U.S. government, military, media, and academic establishment.

To carry off this implicit agreement, the Mexican government adopted institutional forms in political areas having pronounced red-white-and-blue hues. The result is that, through this contrivance, the actual political process is unknown in its fundamentals.

When people say in private, "The president of Mexico decided, for purely egotistic reasons—to divert criticism of his management of the economy—to nationalize the banks in 1982," they are personifying an unknown process. Similarly, when one errs if he comments that the president's successor's taking that decision as an irreversible one shows his political weakness: the authority and the functions of the Office of the President of Mexico are unknown, and there is no reason to believe that, acting as solitary office-holder, he has the power to reverse the nationalization decision.

This agnostic stance is not the majority position in Mexico, where people believe that in his own country the President of Mexico is the most powerful man in the world. Mexicans go to great lengths to convince the U.S. media and the Mexican people that decisions announced by the President are indeed those of the President. But explaining major political and economic events in Mexico as the result of "decisions by the President and his closest advisors" is at bottom an inference.

Take the energy sector. Who, really, is in charge? It was said that President López Portillo and Pemex director Díaz Serrano once had a talk about creating an energy ministry. "George," the president said, "I can't let you be head of any new such ministry—you would be taking my job."

Creating a Mexican Department of Energy does not, as President Carter learned, solve political (and perhaps not even administrative) problems. The paradox of the 1983–1988 National Development Plan of the de la Madrid government is that the authority of the energy sector minister is no stronger than the agreements negotiated with other cabinet ministers. The American solution to this problem—appoint a Henry Kissinger to the President's White House staff—simply won't work in Mexico, given that the Mexican State can only operate with one media personality at a time. If no one is de la Madrid's personal staff can be given authority, in public at least, for keeping cabinet ministers in line, it means that de facto decentralization at the cabinet level is the normal state of affairs. Real policy coordination across cabinets is likely, therefore, to be the exception, not the rule.

With regard to the several agencies within the energy ministry, is it realistic to suppose that a newly created ministry will exercise real authority over Pemex and the CFE, two vastly different institutions whose goods and services are alike only in the respect that they can be converted into common equivalent units of heat. That these two organizations should belong under the same administrative and operational authority is an ideological postulate, not an industrial or commercial imperative.

An alternative to this great-man (Napoleonic-style) view of political events in Mexico is hard to articulate, owing to the lack of data establishing the shape and function of political authority in Mexico. Only one thing is certain: no one in Mexico believes the explanation that political decisions reflect the views of voters and that laws reflect the deliberations of legislators.

Emergence of Mexican Resource Nationalism

Mexico's attitude toward the outside world has been described by Mexican writers such as Octavio Paz as suspicious, nationalistic, and introverted. Thus, the outside supplier, trader, or investor in Mexico eventually is forced into generalizations about Mexico's history, culture, and people. But Mexico is one of those countries of which it is said a person can stay a week and write a book; stay a month and write an article; stay a year and write nothing.

What it comes down to is this: with so much government and media rhetoric about the powers, rights, and duties of the State, one must ascertain how much is real. Either the Mexicans are among the most State-fearing people on earth, for whom the State exists as a mystical, corporate-like moral force and authority, or else this stunning rainbow of political, economic, and juridical ideas exists for another purpose—and the Mexicans, consciously or unconsciously, know what that purpose is.

As with most developing nations, Mexico has ambivalent feelings toward the U.S. Mexican humor recognizes this ambivalence in several well-known expressions: "Poor Mexico—so far from God, so close to the United States" or "Thank Heavens we've got the United States—if they didn't exist, we would have had to invent them."

While these expressions point out the Mexican ambivalence toward the U.S., they fail to even hint at the natural comparative trading and investment advantages that geographic proximity would seem to offer—independent of any consideration of Mexico's potential role as a major supplier of crude oil to the U.S. But to understand Mexican oil policies, an adequate grasp of several trend lines in Mexican history is needed.

Mexico's Spanish heritage

In 1492 Spain came to the surface of European life after 700 years of foreign domination by Arabs. In that year the last Moorish outpost was overrun, a victory that was both military and religious. But the suspicion of Spanish Catholics toward foreigners did not stop with the expulsion of the Moslems. Spanish Jews were expelled as well, an act that cost Spain much of her managerial talent and middle class.

In the same year other far-reaching events occurred: a Spanish pope was elected, and the first Spanish grammar was published. Equipped with an army of veterans, a militant evangelical clergy, and a centralized literate bureaucracy, Spain emerged as the first modern, corporate state of Europe. This ideological, totalitarian model of the state was Spain's first export.

For Spain the discovery of the West Indies represented but one of several new opportunities for conquest and colonization. Given the doctrine of the universality of the Christian religion, a bureaucracy headed by rulers committed to the expansion of Spanish institutions, civil and religious, it is no wonder that Spain made such a great effort—first to conquer, then to seal off her American colonies. In the process, Spain came to see herself as having a unique international role and mission. Meanwhile, Spanish jurists such as Francisco de Vitoria systematized their views on the duties and relationships among independent states and in so doing laid the foundations of modern international law. In some ways, 20th-century Mexico is a mirror image of 16th-century Spain.

A transcendental event in Mexican history was the forced marriage of meso-American and European destinies. Aztec society was ruled by a leader with unified powers: military, religious, political, and economic. The Aztecs at that time were nervous and fearful; a once-banished, half-historical, half-mythological figure, Quetzalcoatl, was prophesied to return to rule Mexico. The curious detail was that in popular legend Quetzalcoatl was white and bearded. At that precise moment, Spain entered.

Though shocked by the sight of Spanish firearms and horses, the Aztecs were not at all surprised by the arrival of these emissaries of Quetzalcoatl. Call it an accident of history that a few Spanish sailors who had shipwrecked off Yucatan several years before had befriended an Aztec woman of noble birth who spoke the Mayan language. When rescued by Hernando Cortez, the sailors, who had learned the rudiments of Mayan, took the woman, Malinche, with them. Malinche became the mistress and interpreter of Cortez who, from Spanish to Mayan to Aztec, could communicate with the Aztec chief of state, the gloomy Montezuma.

The Aztec state, though only under a minor threat to its national security—a few hundred Spaniards against tens of thousands of Aztec soldiers—was defeated by the confusion and disagreement over the identity and purpose of the foreigners in Mexico. Having defeated the Aztecs, the Spaniards built a very Aztec-like state.

From 1535 to 1821 Mexico was ruled by a Spanish-appointed viceroy, and during this long period Mexico learned its enduring political habits. The viceroy, much like the modern-day president of Mexico, had unified powers: he was the chief executive, the chief justice, commander in chief, and civil head of the Mexican church. There was no person nor political body below him to whom he was answerable.

When independence from Spain occurred, Mexico began its long experiment with viceregal-cum-constitutional government: political institutions with viceregal substance and constitutional dress were invented, modified, and typically discarded. The style of dress varied among *caudillos*. There were the men-on-horseback types exemplified by Emperors Iturbide and Maximilian and General Santa Anna. There were Jacobin reformers fired by a strange Mexican anticlerical passion—men such as Valentin Gomez Farías and Benito Juarez, who were tireless in their search for a political framework that would hold Mexican society together. But the Texan revolt took place before that framework had been perfected.

The uprising by Mexico's northeasternmost province, which in its turn brought on the war with the U.S., forced Mexico to come to terms with long-range political strategy. Up until then, Mexican political theorists had assumed that title to Spanish lands automatically passed to the new Mexican state and acted as if a probate court had awarded what is now Central America and the western half of the U.S. to a few hundred political activists in Mexico City and outlying towns.

In the early 1820s Mexico adopted a policy favoring foreign immigration into its northern provinces. When, by the mid-1830s, the Europeans and Americans in Texas became unmanageable, Mexico's presidents-on-horseback tried to subdue them by force and, in one memorable instance at the Alamo, succeeded. When Texas, whose political independence Mexico never recognized (from 1836 to 1845), was finally annexed by the U.S. in 1845, a stage was set for conflict. The American position was clearly that Texan territorial disputes now were matters of U.S. government cognizance. Mexico's house, meanwhile, was divided. Most of the tiny minority of the politically active population sided with the position that Texas still belonged to Mexico and that the presence of U.S. military forces in Texas was tantamount to an act of war.

Mexico City's closed society was burst open by the military invasion of the U.S. Army. General Winfield Scott occupied the capital of Mexico on September 15, 1847. Shortly thereafter President Santa Anna resigned, and the Mexican government on a national scale ceased to exist.

With the fighting over, Mexican politics divided into two camps. One camp said that Mexico, if it was ever going to modernize itself, must model itself on the democratic institutions of the U.S. The presence of the U.S. Army in Mexico was, therefore, seen as a blessing in disguise, for only the U.S. Army could break the privileges and abuses of the Church (said to own over half the national territory). The other camp believed that Mexico must remain in the hands of the established families, descendents of the Spanish conquistadors, who had controlled the Church and army for 300 years.

Nicholas Trist, who studied law under Thomas Jefferson, made an unauthorized treaty with the proclerical, antidemocratic party, thus guaranteeing the rule of conservatism in Mexico. In retrospect some Mexican observers have said that the U.S. invasion was the catalyst that crystallized Mexican nationalism, giving it an indelible, anti-American signature.

Conservative Mexico reflected on the unseemly enthusiasm for American political institutions on the part of public figures such as Miguel Lerdo de Tejada, Manuel Crescencio Rejon, and Franciso Suarez Iriarte, the latter, head of the pro-American Municipal Assembly of Mexico City that briefly held power during General Scott's occupation rule in the winter of 1847–1848. Just as Mexican proclerical conservatives had won power from Spain in 1821 under the guise of liberalism, so Mexican anticlerical conservatives of 1857 and succeeding generations would devise constitutional rights, freedoms, and Yankee-style political mechanisms as the best defense against renewed, armed, pro-Americanism.

The Porfirian vision, 1876–1910

Porfirio Díaz was a standard, rightist strongman of Latin America: unbending on the issue of domestic stability and control, favorable to foreign investment, and keen on utilizing the latest in technology and management concepts available in Europe and America. Under Díaz's 35 years of rule, Mexico made more economic progress than it had during the previous century.

Díaz opened the country to British, American, and French companies seeking investment markets and raw materials. One inducement was the mining law of 1893, which gave subsoil mineral rights to property owners. This law was a major break with Spanish law, under which subsoil wealth belonged to the State. The presence of foreigners and foreign interests in Mexico became so pronounced under Díaz that it expressed itself in that infallible index of Mexican attitudes, political irony. "Mexico, mother of foreigners, stepmother of Mexicans" went the expression of the day.

Since Díaz's day, Mexican church dogma in matters of national history holds that the era of Porfirio was an unfortunate anomaly, especially so in its illegal precepts pertaining to subsoil mineral rights. While Díaz's innovations were indeed a break with Spanish tradition, Mexico, since its political divorce from Spain in 1821, was under no obligation whatsoever to perpetuate Spanish legal notions or practices. As a sovereign state Mexico was free to experiment. Although, in a given instance, the results were not as favorable as had been expected, that was not an argument that the experiments were philosophically invalid because they contravened colonial legislation of centuries past.

Private sector management

The petroleum crisis of the 1970s drove home the point that energy is an internationally priced commodity and that an abrupt change in its cost or management profoundly affects national economic planning and international relations. The revolution in Iran was a case in point, and Mexico is sometimes mentioned in the same breath.

At a time when the energy dimension of U.S.-Mexican economic relations is taking on increasing importance, one should recall that the first oil crisis in Mexican-American relations took place on the stage of a management revolution in Mexico. The impending crisis became evident in 1917, when American investors and government leaders learned that the planners of the new Mexican constitution had rewritten the rules for the management of the petroleum industry.

Article 27 furnished the revolutionary design for the reform of agrarian and subsoil wealth. Its purpose was not merely the redistribution of wealth but its reorganization as well. Large agrarian estates would be reduced in size or expropriated so the *campesino*, Mexico's peasant, might have land. Owners of petroleum enterprises and contracts would be permitted to carry on their operations only under the regulatory eye of new federal legislation.

This early history of U.S.-Mexican energy diplomacy adumbrates much of the current discussion of natural gas, oil, trade, and immigration. During the period 1917–1938 the U.S. government invoked three principles of law and diplomacy—linkage, good neighborliness, and a theory of just compensation—in the hope of persuading the Mexican government to adopt a policy favorable to U.S. property holders in Mexico.

The crisis provoked by the new Mexican document anticipated the crisis that would surround the entry of the American government into the energy-management field with the creation of the Tennessee Valley Authority (TVA) in 1933. During the period 1917–1927 Mexicans faced the hostility that New Dealers who supported the TVA would meet from conservative spokesmen of industry and government. An important difference between Article 27 and the TVA Act was that the first was a constitutional precept, while the second was direct, enabling legislation. The first was a game plan, the second a fait accompli. To have practical application, Article 27 required corresponding legislation. This point was precisely where American attention was concentrated.

The Bucareli Conference of 1923 sought a compromise between the American conservative position and the Mexican liberal constitution. Americans, vigorously insisting upon the validity of their interpretation of Mexican constitutional law, had denied that an abrupt (revolutionary) change had taken place in Mexico's legal fabric. Since civilization itself depended upon the continuity of legal rights, property rights that were acquired by law could not be destroyed by law. Hence, the new Mexican government needed to recognize American property rights or its refusal would justify a denial of diplomatic recognition.

American investors rejected all arguments for legislative change that would damage property rights. For them, the question was never the renewal of rights but only the recognition of rights. These investors exerted tremendous pressure on the American government for persuasive diplomacy—and armed intervention, if necessary—to preserve their pre-Revolutionary rights in Mexico.

Oil in Mexico's new constitution

For its part, the Mexican constitution had simply said that land and natural resources belonged to the nation originally. But what legal or practical effect was the term "originally" to have? Mexican and American leaders never agreed upon an answer, and the diplomatic milestones of 1917–1938 in retrospect were diplomatic disguises of the failure to reach an agreement. During this period the U.S. government naturally was concerned about the protection of American rights in agrarian and petroleum interests.

The official American position was that 1) all rights legally acquired by American citizens prior to May 1, 1917, the date that the constitution went into force, belonged to the individual or corporate property holder; and that 2) American property holders of this class were exempt from the regulatory provisions that might emanate from the Constitution of 1917 to the extent that the new codes were in conflict with the legal system that existed at the time the property was acquired. In short, American property rights were deemed unaffected by the social and constitutional revolution in Mexico.

The position of the Mexican constitution-makers was naturally quite different. Their position was that the legal system in force prior to the Revolution had been a system of notorious social injustice. The natural and human resources of the nation had been put in the service of economic and international elites with no regard for the moral or material development of the Mexican people. The Constitution of 1917 was to be the corrective, a national plan for the development of a just, equitable, self-modernizing society.

In order that this newly defined national purpose might be achieved, it was necessary that all legal rights acquired prior to the date the plan went into effect be carefully reviewed. National policy required that such rights be either cancelled or renewed, depending on the benefit in kind and amount available to the nation by the exercise of a right in question.

Early oil policies of Mexico

On April 2, 1921, President Alvaro Obregon, following the policies of his revolutionary predecessors, announced to the foreign press that the future regulations for the implementation of the controversial Article 27 would be governed by a spirit of equity. There would be no regulation of a confiscatory character nor one that would permit a retroactive interpretation.

The American response took the form of a proposed treaty of amity and commerce, the terms of which, having been approved by President Warren G. Harding, were delivered personally to General Obregon on May 27, 1921. On June 4, 1921, Foreign Minister Alberto J. Pani definitively rejected the American suggestion that diplomatic recognition occur simultaneously with the signing of a treaty that would have as its purpose the protection of American property rights. As Pani and his successor, Aaron Saenz, would have to insist repeatedly during the next two years, Mexico would never accept conditional recognition.

Frustrated, U.S. Secretary of State Charles E. Hughes explained to the press on June 7, 1921, that the major question about Mexico held by the U.S. was how the government could safeguard American property against confiscation. A retroactive application of Article 27, he warned, would be seen as an international wrong to which the U.S. could not consent. On June 11, 1921, the Mexican general asked his American counterpart what additional assurances were needed from the Mexican government to demonstrate that Mexico had both the ability and willingness to comply with its international obligations.

On August 30, 1921, the Supreme Court of Mexico reported a pivotal finding in the case of the Texas Oil Company: Ownership of petroleum rights under the 1909 mining code was not vested until some positive act was undertaken to possess the subsoil deposits. The nationalization program, the court explained, would only affect property on which no such positive act had been carried out prior to May 1, 1917.

In his first annual message to congress, General Obregon pointed to the concurrence of the three branches of government in the view that no retroactive application of Article 27 would be permitted under the constitution. Before the same body a year later, President Obregon reported that the U.S. government would continue to withhold diplomatic recognition until such time that it received adequate guarantees covering the property rights of Americans that had been acquired legally prior to the Constitution of 1917. He added that while his administration had rejected the American offer of a treaty because it had implied conditional recognition, it had proposed the signing of two claims conventions.

In his instructions of May 8, 1923, to the two executive agents who would spend their summer in Mexico City in daily conference at 85 Bucareli Street, the American foreign minister pointed out that although diplomatic recognition was a subordinate question, the two claims conventions could not be signed without it. Diplomatic relations would be renewed, he cautioned, only if the American government were to receive substantial assurances that the Mexican government recognized and would discharge its international obligations—among them adequate protection of the valid titles of American citizens. Secretary Hughes added that his government did not regard stability as the sole criterion for recognition; what counted more heavily was the ability and willingness of a government to discharge international obligations without recourse to confiscation or repudiation.

Although the American minister had insisted for two years that an adequate assurance could be obtained only by means of a treaty, Hughes left the form of the assurance up to the commissioners. The necessary element in any assurance, however, would have to consist of proof that the Obregon administration was pledged to a policy that would relieve the American government of its doubts of recent years.

Since Mexican law had established the precedent that no application of Article 27 would affect properties on which a positive act of exploitation or exploration had taken place, the outstanding point of difference between the two governments in this area was the application of the nationalization program to lands that were being held in reserve by petroleum interests and upon which therefore no operations had begun.

The American government had rejected the Mexican distinction between vested rights and mere speculation, a distinction that would permit, according to Mexican constitutional theorists, the expropriation of speculative petroleum lands. For this reason Foreign Minister Hughes insisted that the American commissioners obtain satisfactory assurances against confiscation of all subsoil interests in lands owned by American citizens prior to May 1, 1917.

With respect to agrarian properties owned by American citizens, Secretary Hughes admitted that his government was in full sympathy with Mexico's attempt to redistribute large holdings in order to meet the natural demands of the Mexican people; nevertheless, his government insisted upon the restoration or just compensation of American farmlands that had been seized. Compensation, moreover, could not take the form of worthless state bonds but had to be in cash or its equivalent.

On August 24, 1923, Foreign Minister Pani wired a message to his American counterpart. The official announcement that their two governments had decided to renew diplomatic relations would have to be made public no later than August 31, 1923. Pani proposed that the text of their simultaneous statements to the press explain this decision as the favorable result of the Mexican-American conferences that had taken place in Mexico City.

On September 1, 1923, General Obregon explained to Mexico's congress that the decision to renew diplomatic relations, which had been announced officially by both chancellories, had been taken as a result of the direct, informal conversations that had taken place between representatives of the presidents of the two countries. From May 14 to August 15, 1923, talks were held between the general's spokesmen, Ramon Ross and Fernando Gonzalez Roa, and those of President Harding, Charles B. Warren and John B. Payne. The exchange of impressions and views was achieved. The Mexicans convinced their American colleagues that the goal of their government, the economic and moral development of the Mexican people, would be carried out in accordance with the constitution by programs that would also encourage the prosperous growth of foreign interests in Mexico. In no way, the Mexican president emphasized, was American recognition sought by treaties or agreements that would be in conflict with either Mexican law, the norms of international law, or the standards of Mexican dignity and sovereignty. The conventions that would be signed, rather, were those that had been proposed by the Mexican chancellory in the informal letter of November 19, 1921, to the American embassy—proposals, General Obregon recalled, that he had already discussed with Congress in his annual message of September 1, 1922.

The first convention would eststablish a mixed commission that would review the claims that an American citizen might bring against the Mexican government for damages stemming from the revolution. The second convention would create a mixed commission that would judge upon the claims that either a Mexican or an American citizen might have against the government of the other for damages that occurred since the signing of the claims convention of July 4, 1868, excluding the period embraced by the first convention (November 20, 1910 to May 31, 1920).

Since the formal conventions were designed to clean up the past by furnishing mechanisms to deal with claims for damages, what kind of informal or unofficial assurances about the future did the Mexican commissioners offer that would lead the stubbornly legalistic American government to recognize the Obregon administration? What part would the unofficial understanding of 1923 play in U.S.-Mexican diplomacy during the years 1924–1938?

The informal conversations of 1923, which one student of Mexican-American affairs called the Recognition Conference, settled the nerves of the American government. Americans holding legal rights in agrarian or petroleum interests would be exempt from any retroactive or confiscatory application of Article 27. All sides anxiously awaited the promulgation of the Organic Law that would specify the ways the article would be applied. Interest in this question was so high that on November 17, 1925, Secretary of State Frank B. Kellogg delivered an aide-memoire for the benefit of Mexican lawmakers.

When the regulatory act of December 26, 1925, finally arrived, a wave of indignation among foreign governments and petroleum interests coursed. The new law required that all petroleum operations be expressly authorized by the federal executive. A confirmatory concession that would extend for 50 years beyond the date of initial operations would be issued to persons who acquired legal rights to subsoil deposits prior to May 1, 1917, provided that an overt or positive act had been undertaken for the exploitation of petroleum.

American petroleum owners had remained uneasy about the constitutional status of their reserved oil lands since the enunciation of the doctrine of positive acts by the Mexican supreme court on August 30, 1921. Since the new regulatory act could deprive owners of 80–90% of their total real estate holdings, and since the principal point of contention—which they thought had been settled in 1923—was the status of land reserves upon which no work had begun prior to May 1, 1917, American owners protested that the new law violated the agreements of 1923. Their protest, in turn, was registered officially by the American ambassador. The Mexican chancellor tried to deflect the complaint by pointing out that in 1923 the Mexican commissioners had promised that preferential rights would be extended only for a limited time to persons who had not performed positive acts with respect to their petroleum properties.

Naturally, the American government in Washington wondered what had become of the binding force of the general understanding reached in 1923. Foreign Minister Aaron Saenz replied that the Mexican government recognized no force in that understanding that might be seen as the equivalent force of a treaty or constitutional precept. Saenz suggested that the points exchanged in 1923 should be seen, not as a binding international agreement, but as unilateral policy outlines that had been designed for the development of plans and programs during the Obregon administration. Americans should feel satisfied under either interpretation, the Mexican chancellor chided, for Mexican policy in the main had followed those outlines set forth in 1923.

His American counterpart warned that relations between the two countries would take on an extremely critical character were laws enacted and enforced in Mexico that violated either the fundamental principles of international law or the terms and conditions of the understanding reached in 1923. The U.S., Secretary Kellogg emphasized, expected the Mexican government to act in accordance with the true intent and purpose of the negotiations of 1923.

In American eyes, what was the true intent and purpose? The Mexican government in 1923 urgently had sought the renewal of official relations between the two countries and, as a consequence of the conference, obtained diplomatic recognition. The Americans understood that recognition had been granted in exchange for a Mexican commitment to respect the acquired property rights held by American citizens. For this reason, Secretary Kellogg reiterated this point in his letter of October 30, 1926, to the Mexican chancellory: his government expected that no action would be taken pursuant to Mexican laws or regulations that would deprive American property holders of the full exercise of their rights.

In summary, the American viewpoint in 1926 was that the negotiations of 1923 had produced two political conditions: one was recognition of the Obregon administration and the other was a *fuero,* or special status, in favor of the American property holder in Mexico, which put his economic pursuit beyond the pale of Article 27. The Mexican viewpoint of 1926 was that the points discussed between the executive agents during the conferences of 1923 could not be considered binding on the legislative bodies of either nation and certainly no executive agreement could modify constitutional precepts.

The 1923 agreements became the topic of indirect discussion following the nationalization of the petroleum industry in 1938. Several officers of the Department of State lectured the Mexican ambassador upon the reciprocal character of the Good Neighbor Policy: The U.S. government was entitled to respect for obligations due it under international law, which included respect for obligations due its citizens.

The American government, the argument went, had always recognized that a sovereign nation has the right, under international law, to expropriate—dependent upon the willingness and ability of that nation to make appropriate compensation at the moment of expropriation. Therefore, the Mexican government would be acting reasonably in the eyes of the American government were it to make no further expropriations unless they were accompanied by effective compensation. The U.S. recognized that Mexican agrarian programs were designed to achieve the social betterment of the Mexian people; yet such programs could not be carried out at the expense of American citizens. The Good Neighbor Policy, in short, could not be unilateral.

This line of legalistic moralizing on the part of the American chancellory made a poor impression in Mexico. Foreign Minister Eduardo Hay denied the existence of any universally accepted theory in international law or practice that required payment of immediate or deferred compensation for expropriations made for the redistribution of land. His government, he insisted, was following a course of action that sought justice and the improvement of a whole people, not a course that sought to protect the pecuniary interests of a few individuals.

The American chancellor rejected the Mexican position: the doctrine of just compensation for property taken originated before the advance of international law and no government could expropriate private property, for whatever purpose, without providing prompt, adequate, effective payment. The date of the American reply, August 22, 1938, was just 15 years after the close of the Bucareli conference. The American argument was based—not on the supposed binding force of the 1923 agreements nor on the supposed reciprocal character of the Good Neighbor Policy nor even on a precept of international law—but on the supposed primordial doctrine of just compensation.

The American argument during those 15 years was contradictory. While recognizing that, independently of any provision of the 1917 constitution, the Mexican government had every right under international law to expropriate the entire petroleum industry, the U.S. government insisted that American property rights acquired before May 1, 1917, were free from any retroactive application of Article 27.

The explanation for this double standard was found in the American attitude toward the special case of Mexico: Since the Mexican Revolution did not pretend to be of the Bolshevik variety, there could be no valid Mexican claim to a major change in the system of private property. For the Americans it was as if, by definition, the Mexican Revolution was not revolutionary. This attitude explains why the Americans so self-righteously defended property rights acquired during the Porfiriato, the period of Porfirio Díaz' rule. In the American legal mind there had been no revolution in Mexico.

The Mexican argument during those 15 years was also contradictory: by entering into a 10-year discussion of the question of whether to apply Article 27 retroactively, Mexican leadership lost sight of the real issues. Article 27 said that real property and subsoil wealth belonged to the nation *originally*. The deceptive American query, "Do you mean *after* May 1, 1917," might have been rejected as irrelevant from the start. Because it gave foreign governments and petroleum interests legal ground upon which they could challenge the Mexican government in its efforts toward an effective application of Article 27, the homemade Mexican doctrine of positive acts became more of a handicap than a help. A better technical solution to the problem of the recognition of the existing preferential rights of petroleum owners (and agrarian property holders as well) would have been that of acknowledging, *ex gratia* (for free), rights of a character that the Mexican government would have been free to specify.

In the Special Claims Convention of September 10, 1923, Mexico waived the generally accepted rules and principles of international law in favor of accepting, *ex gratia,* a moral responsibility to make full indemnification for damages suffered by Americans during the revolution. Similarly, Mexico could have offered, *ex gratia,* petroleum concessions to persons or companies that had begun operations prior to May 1, 1917.

By advancing the concept of confirmatory concession, Mexico affirmed that legal rights to subsoil wealth had survived from the old regime. A more revolutionary position would have been to say that all subsoil rights ended as of May 1, 1917. Probably such a position would have been too threatening to American political sensitivities. Mexico had already suffered two American military interventions, and the Obregon administration badly needed U.S. recognition, money, and arms.

For reasons of its own, the U.S. government was determined to play the game of protecting the rights of its citizens abroad. The Americans tacitly asked for assurances in exchange for nonintervention, and successive Mexican presidents from Carranza to Cárdenas were willing to pay that price. In playing the American legal game, however, Mexican diplomats were also forced to act as if there had been no revolution in Mexico.

Apart from the narrow place of the Bucareli conference in the history of U.S.-Mexican relations, Mexicans had two questions of greater transcendence:

were the procedures followed during and subsequent to the conference in violation of the Constitution of 1917, and did the results of the conference retard or accelerate the implementation of Article 27?

The answer to the first question is negative if one takes the view of foreign ministers Pani, Saenz, and Castillo Najera that the conference was a direct, informal exchange of impressions and views and not an international agreement. The answer is affirmative if one takes the view of foreign ministers Hughes, Kellogg, and Hull that the U.S. government granted *de jure* (by law) recognition to the administration of General Obregon only because of the strong assurances that the American government had received with respect to the protection of American property rights in Mexico.

By adhering to this second view, one could infer that constitutional procedure was violated because the Constitution states that the Senate has the sole power to approve treaties and diplomatic conventions. Since the exchange of executive views, called the Unofficial Pact, was approved by both presidents without the advice and consent of either Senate, constitutional irregularities occurred in both countries.

With respect to the second question, one must explain the significance of the 15 years that elapsed between the conference of 1923 and the expropriation of 1938. Can the failure to achieve adequate governmental regulation of the petroleum industry in accordance with Article 27 be ascribed to the intentional nonapplication of the article by the Mexican executive branch because of a prior commitment of a supraconstitutional character to the Unofficial Pact of 1923?

Since the Constitution required regulation, not expropriation, no case can be made for an earlier date of expropriation. The deterioration of the political and economic relationships between the Mexican government and the petroleum industry can be explained by reference to the defiant attitude of the industry that sprang from an unswerving adherence to the proposition that although original ownership of subsoil wealth might have belonged to the Nation, by the mining codes of 1884, 1892, and 1909, permanent ownership had been transferred to private interests.

The uncooperative attitude and behavior of the petroleum industry would have continued independently of the results of the conference of 1923. The oilmen never understood that the willingness on the part of the Mexican government in 1923 to acknowledge preferential rights to subsoil deposits was in no way a willingness to acknowledge direct, private ownership.

By its insistence on legalistic formulas and solutions and by its traditional, counterrevolutionary perspective, the U.S. government indirectly gave moral support to beleaguered American property owners. The Mexican government was hampered in 1926 and 1927 in its attempt to enact regulatory legislation for Article 27 because of the harassment by American authorities who believed that Mexico had been bound legally by the results of the conference of 1923.

Because Mexican governments had made the mistake of letting the Americans hold their erroneous view that there had been no legal revolution in Mexico, Mexican diplomats repeatedly found themselves in the embarrassing position of pretending that the minutes of the Bucareli conference had legal force. In fact, those minutes had no legal force in national or international law, and Mexican diplomats never doubted this point. But had this point been made clear in 1923, diplomatic recognition would have been doubtful.

Worker-technician management of the oil sector

Citizens on both sides of the Rio Grande sometimes forget that American firms were not the only ones affected by the expropriation of foreign oil company holdings. English and Dutch interests represented 70.5% of the foreign petroleum industry in Mexico. After the ruling of the Mexican Supreme Court in favor of the workers, the Anglo-Dutch companies announced an international campaign against Mexico and withdrew their funds from Mexican banks. Britain severed relations with Mexico, an action the U.S. government refused to take.

Some observers suggest that the U.S. government and even U.S. oil companies supported the expropriation. Evidence of this alleged collusion exists in letters from Cardenas to President Franklin D. Roosevelt, thanking him for his support of the concept of national sovereignty; Roosevelt's recommendation to the companies to accept Mexico's right to expropriate; Roosevelt's role in repairing Anglo-Mexican relations after the expropriation; and, most curious of all, the fact that long after expropriation Mexico became a net importer of American technology and petroleum products.

Therefore, according to this reasoning, the Americans knew from the start that expropriation would eliminate English competition and that Mexico could not operate on its own resources, leaving Mexico no alternative but to turn to the U.S. for technical and financial help.[1] Given the U.S. recession and an oversupply of oil worldwide, U.S. oil companies were willing to sacrifice their Mexican interests in order to foil the British and strike a blow at Mexico's export capacity.

Such an interpretation ignores the many legal and extralegal actions and the vituperative media attacks launched by U.S. oil companies in retaliation for expropriation as well as Washington's initial cutoff of purchases from Mexico. This example of a wayward view of the 1938 expropriation only throws light on that period of the Mexican oil industry in that it points out the tendency to focus

[1]George Philip, *Oil and Politics in Latin America,* p. 331, prints the text of a state department memorandum that suggests that Pemex's decision not to readmit foreign oil producers in 1947 was influenced by Everett de Golyer. Golyer argued that Pemex, with the assistance of U.S. oil-field service firms, could operate and expand Mexico's petroleum industry. If so, the marketing ploy succeeded—necessarily at the expense of the integrated producers.

on the nefarious motives of the U.S. actors and let the Mexican actors assume the mantle of heroes of economic nationalism. In fact, the expropriation touched off a battle between the petroleum workers and the government that has left a lasting mark on Pemex management policies.

Cardenas was driven to the expropriation act by the companies' resistance to labor's demands and finally their refusal to abide by the ruling of Mexican courts in favor of the petroleum worker's union, STPRM (Sindicato de Trabajadores Petroleros de la República Mexicana). The petroleum workers, or *petroleros,* already had the best pay and most generous benefits of any major group of workers in Mexico, and it whetted their appetites for more.

After the expropriation of March 18, 1938, workers and officials of STPRM stepped in to fill the vacancies left by the departing foreigners. Many of them worked long hours without extra compensation and improvised ways to circumvent production snags caused by the often decrepit equipment left behind and the lack of spare parts. Not surprisingly, the members of STPRM took pride in their role in the expropriation and in maintaining operations after foreign personnel left. They felt a proprietary interest in the industry and even a paternal pride in the birth of the new nationalized oil industry.

The ambitious aspirations of STPRM ran into stiff opposition from the other parent of the nationalized oil industry, President Lazaro Cárdenas. STPRM espoused a belief in syndicalism as the way to run the economy and fought for control of the industry, citing the railroad workers' assumption of control of the nationalized railroads in 1937. Given the abysmal performance of the railroads since nationalization, this was hardly a convincing precedent from the government's point of view. The political dispute over whether the expropriation was done for the benefit of the *petroleros* or for that of all Mexicans, for syndicalist or socialist aims, is of less concern than how this period affected Pemex management.

Pemex did not suddenly appear the day after March 18, 1938. It descended from the National Petroleum Administration (NPA), chartered by the government in 1925 as a public agency to compete with foreign companies in production and refining. In 1934, NPA's functions were transferred to a semiprivate organization, Petróleos de Mexico (Petromex), in which foreign stock ownership was forbidden and the government reserved a minimum of 50% ownership for itself.

In 1937, anticipating nationalization of the industry, the executive department created a government corporation to assume the property and functions of Petromex. On the day of expropriation, this corporation, the General Administration of National Petroleum (GANP), had a reported production of 16,000 b/d. GANP, its predecessors, and the foreign companies were the training ground for those workers and union leaders who stepped into the vacancies left by the departing foreigners.

The government had assured itself of a cadre of trained workers through a series of legal measures. The Labor Law of 1931 required that 90% of the workers in any establishment be Mexican. An immigration ruling in 1933 mandated that individual contracts be drawn up for every technician imported and required the company bringing the technician to Mexico to train a Mexican replacement by the expiration date of the contract. So the industry Mexico inherited consisted of 18 independent company organizations formed in a competitive environment. Taken as a whole, each organization duplicated the functions of the others, representing too much redundancy for an integrated and rational state-run industry.

Labor gained a valuable foothold during the period immediately following expropriation because of the government's provisional solution to this administrative problem. Local-level administration of the industry was assigned to 18 boards (Consejos de Administración) established by divisions of the STPRM, which was organized locally on a company basis. Nationally, the government set up an administrative council of nine members, five from executive departments of the government, one from GANP, and three from STPRM. Initially, the local boards cooperated well with the national council, primarily because they had almost complete autonomy and were lobbying for permanent syndicate control of the oil industry.

With the permanent plan of the industry's organization enacted on June 7, 1938, STPRM's hopes of complete control were lost. The law established two new public agencies: Petróleos Mexicanos (Pemex) to handle exploration, production, and refining and Distribuidora de Pemex to market petroleum domestically and abroad. Pemex management consisted of a nine-member board of directors, with the same distribution of representatives as the provisional administrative council of the oil industry. Distribuidora de Pemex had a five-member board of directors without any representation from STPRM. In its charter, Pemex was given the principal goal—not of profit—but of achieving social rather than speculative accomplishments. Presumably, "speculative" referred to exploratory and production work that entailed high levels of risk to the capital invested. However, the government, including the self-professed communist Jesús Silva Herzog, felt Pemex should earn enough to cover operating costs and have enough surplus to build schools and hospitals for workers, meet the development costs of the industry, and earn a profit for the government. Prices, though, were not geared toward maximizing profits.

Pemex may have been an autonomous government agency, but the executive department of the government exercised strict price controls over the industry from the beginning. The July 1981 firing of Pemex president Jorge Díaz Serrano showed that the executive branch has continued to guard its pricing prerogatives jealously.

The leaders of STPRM did not regard Pemex as a commercial enterprise but as a vehicle for carrying worker benefts and wages to higher levels and for instituting national economic planning. The STPRM operated from a powerful vantage point. As the foreign companies withdrew, 1) *petroleros* and STPRM officers stepped in to fill the most important job vacancies in the company structures; 2) the local STPRM divisions ran the local boards, both under the provisional management scheme and after the establishment of Pemex and Distribuidora de Pemex; and 3) labor had tremendous prestige and popular support for its role in the nationalization.

When the government tried to reorganize the industry to eliminate the duplication of effort and expense inherent in the parallel structure of the expropriated firms, STPRM balked, fearful of losing jobs. In 1939 after a bitter fight, management succeeded in centralizing accounting, legal, medical, and engineering services by setting up unified departments in Pemex. Labor's opposition did prevent the absorption by Pemex of the GANP, however.

Even more fierce resistance greeted management's proposal to eliminate Distribuidora in order to centralize control of domestic marketing, which labor still controlled at the local level under Distribuidora. Pemex finally centralized retail agencies in each city by the end of 1939 but took even longer to centralize the bulk distributing plants. STPRM was strong enough, however, to block the elimination of Distribuidora, and all personnel displaced by the centralization of sales were not laid off but were employed to set up new agencies, mostly on the northern border. Pemex's reputation for featherbedding was born during this fractious period.[2]

By 1940 the industry was in a woeful state. In April 1938 oil employed 15,895 workers; by January 1940 the number had reached 19,316, a 22% jump. The first president of Pemex, Vicente Cortés Herrera, exhorted the STPRM to "save the industry" from labor's excesses. He noted that the industry's budget had grown by 22 million pesos during the first 22 months after expropriation, but the benefits to individual workers had not improved because of padding the workforce. Despite the larger number of employees, production had decreased, indicating declining worker productivity.

Even labor's staunch ally, President Cárdenas, leveled charges at STPRM. In April 1940, he said, "A lack of understanding of new conditions in the industry on the part of a small nucleus of workers, among whom are several labor leaders, has led to serious difficulties, unlimited demands, and indirectly a constant rise in expenses, as well as an increase in the unproductive burdens weighing on the industry, with inevitable repercussions for all who work in it."[3]

---

[2]J. Richard Powell, *The Mexican Petroleum Industry* (Berkeley: University of California Press, 1956), pp. 124–156.

[3]Carlos Alvear Acevedo, *Lázaro Cárdenas, el hombre y el mito* (Mexico City: Editorial Jus, 1972), p. 454.

By August of that year, Cárdenas had intervened in the reorganization dispute. Both Distribuidora and GANP were abolished and their functions transferred to Pemex. Even though this was a blow to STPRM's ambitions, the organization did receive some compensation: the position on the board of directors formerly held by the now defunct GANP was awarded to STPRM, giving the syndicate four representatives to the government's five.

The acrimonious series of disputes between government and STPRM in the years immediately following expropriation established a competitive and sometimes antagonistic relationship between the two that has lasted up to the present. The government still sits in the driver's seat, but labor sits in the passenger's seat with one hand on the wheel. Labor representatives are on Pemex's board of directors, and one former labor leader in Pemex was promoted to the national senate.

Third-Worldism

Mexico's oil production level of 1975 allowed it to resume the status of a crude exporter. The country was ruled by Luis Echeverría, who will not be remembered for his economic policies but for putting into 20th-century legal form the cumulative experience of Aztec, Spanish, and Mexican rulers with regard to the identity and role of foreigners. One law requires 51% Mexican equity and management participation in new foreign investments. Another law regulates the use of foreign patents and trademarks. A third restricts the entry and activities of foreign businessmen in Mexico. These three laws are expressions of the anti-imperialism of the Third World during the 1970s, and indicate the fundamental attitude of Mexicans toward their country.

During the Echeverría period anti-Americanism was being subtly institutionalized, in large measure by educational default. The history of U.S.-Mexican relations was painted in dark colors of conspiracy and intrigue. The seizure of Texas and then of California was cited as proof that the domination of Mexico had always been the policy objective of the U.S. government. Some alleged that this policy of domination was being carried into the 20th century by multinational corporations who come to Mexico greedy to exploit her mineral and human resources.

During this period, the overvaluation of the peso reached its peak, and *malinchismo,* the love of things foreign, appeared in new forms. Mexican consumers asserted their preference for foreign-made goods to Mexican-made ones. In Monterrey there was little retail business of the scale of Mexico City or Guadalajara because shoppers drove to Laredo, San Antonio, or Houston for their clothes, appliances, and anything else they could successfully carry back. Mexicans prefer a U.S.-assembled Ford or Chrysler to one assembled in Mexico.

All of this encouraged a large-scale traffic in contraband goods of all sorts, from color television sets to vitamins. To counteract this infatuation with foreign products, the Mexican government initiated new economic consciousness programs. "'Made in Mexico'—It's Well Made" was one government slogan that, in accordance with the iron laws of Mexican political humor, soon appeared on T-shirts adorned with the outline of a shapely woman.

On March 18, 1980, Petroleum Day in Mexico, President López Portillo announced that Mexico would not joint GATT. For many manufacturers in Mexico, this decision was a relief. It meant that, for example, Mexican consumer electronics companies would not have to compete openly with Sony and Panasonic products being smuggled across the U.S. border.

In November 1980 a new energy plan was announced. Mexico had sold almost all of her crude to the U.S. In the 1970s, however, with the rise of anti-imperialism and anti-Americanism in the Third World, Mexico closed ranks with other developing nations and intensified its courtship of non-American markets and sources of technology and capital. This policy led to tangible results in one industrial area after another. In 1979, Mexico's only non-U.S. crude oil customers were Spain, Israel, and Costa Rica. This pattern dramatically changed in 1980 as Mexico concluded crude contracts with Japan, France, Canada, Sweden, Brazil, India, Jamaica, Nicaragua, and Yugoslavia. In addition, Mexico entered into an agreement with Venezuela to be responsible for providing 80,000 b/d to other Central American and Caribbean countries.[4] Mexico saw the agreement as a step toward implementing, on a regional scale, a proposal for a World Energy Plan presented by President López Portillo to the United Nations on September 27, 1979.

The Energy Program reflected Mexico's continuing national security concerns regarding the U.S., Mexico's largest trading partner. The foreign investment laws of the Echeverría period plus the export guidelines in the Energy Program are derived from a determination to anticipate and define the manner and extent of U.S. government and private capital influence in Mexico's economic destiny.

By limiting Mexico's crude exports to 20% of a buyer's total crude imports, Mexico in theory would have some control over the buyer's dependence on Mexico as a source of crude. This in turn would help avoid a situation in which a shortage of crude on the international market might tempt a buyer to exert undue influence on Mexico. The same logic applied to the Energy Program's goal to reduce U.S. allocation to 50% of Mexico's total crude export sales.

---

[4]Signs in early 1983 suggested that this agreement would likely become a casualty of the depressed oil market.

On July 1, 1982, Miguel de la Madrid was formally elected president of Mexico for a 6-year term. His plank: Mexico has always had its share of problems—we will deal with these.

To decipher Mexican history, however, is not to reduce it to mere social science epiphenomena. The U.S. business visitor to Mexico should realize that Mexican attitudes toward foreigners and the outside world go back much farther than anything that happened in U.S.-Mexican relations.

Even so, if Mexican industrial and governmental behavior were to be understood as the dependent variable in a linear equation containing a hundred independent variables, the one labeled "Access to the U.S. Economy" would be weighted far heavier than any other. This U.S. connection probably explains more, in the statistical sense of the term, than the sum of the remaining variables. Mexico's economic destiny in a given presidential cycle depends more on relationships with the U.S. than it does on OPEC's marker crude price.

10

National policy

Energy Program

The Industrial Development Plan, the Global Development Plan, and the Energy Program, which appeared in 1979–1980, set forth a frame of reference in which an annual overall growth rate of 8% in real terms was thought to be attainable under a set of assumed and forecasted conditions.[1]

The basic approach was a modified Keynesian one in which the government considered itself responsible to carry out the necessary capital investments in a range of industries in order to achieve public-interest goals. The basic goal was to utilize finite hydrocarbon resources in such a way at to develop nonhydrocarbon energy sources and nonhydrocarbon exports.

The plan was based on an analysis of the probable investment and productivity levels that private capital, domestic and foreign, would achieve on its own initiative. Where these levels were considered insufficient, public policy stepped in with indirect and direct tools. Indirect tools included a host of incentives for manufacturers and others whose business decisions support the plan's desired results. These incentives included discounts on industrial energy supplies, tax credits, and preferential financing.

The principal tool, however, was direct government investment. Government planners, in collaboration with managers of government-owned firms, developed capital-spending levels by industry sector. To pay for this ambitious development program, Mexico would rely on crude oil and natural gas export earnings. But unlike most oil exporters, Mexico fixed limits on these exports for the 1980s.

[1]1982 saw negative economic growth in Mexico, the rate of which was −.2% according to the central bank but as much as −6% according to other sources. Optimists suggest recovery might occur in 1985.

The Energy Program of November 1980 contained both descriptive and prescriptive components. Alternative scenarios for the energy sector in general and hydrocarbon exports in particular were presented and analyzed. The bottom-line prescription for the 1980s covered both domestic and foreign sales. The growth of domestic demand for hydrocarbons was to be moderated through conservation measures and price policies, generating a savings by 1990 of an oil equivalent of 1.02 million b/d over base-line projections. The regulation of exports was conceived in both absolute and relative terms: exports were to be fixed for the 1980s at 1.5 million b/d of oil and 300 MMcfd of gas. These export levels were considered sufficient to finance an 8%/year growth of GDP in real terms, provided that profits were reinvested in programs that fostered industry and revitalized Mexico's stagnant agriculture.

The Energy Program called for the construction of five new refinery plants that would have doubled Mexico's refinery capacity. It was also foreseen that, as the Gulf of Campeche fields were developed, the percentage of heavy crude oil produced would increase. The program mentioned no special efforts to market this larger volume of heavy crude; on the contrary, it stated that Mexico's refinery policy will be geared toward lighter petroleum products, despite the mix of crude weights. These and other capital-intensive projects would be required to support a production rate of 4.1 million bo/d in 1990.

One of the constraints affecting Mexico's ability to increase its crude production is the overabundance of natural gas, more than 70% of which is associated with crude deposits. *Energy Détente* (2 October 1980) estimated that if, in 1990, Mexico were to produce 4.8 million bo/d, this would generate 15 bcfd of gas. If the U.S. were in that year to buy all of its imported gas from Mexico, this would still leave a surplus of 6.2 bcfd above domestic requirements. If Mexico is to adhere to the export guideline of 300 MMcfd, the surplus over projected domestic consumption in 1990 will be 11.7 bcfd. The Energy Program's figures differ considerably from those just given; at a crude production level of 3.5 million b/d in 1985 only 4.3 bcfd of gas are foreseen, and at 4.1 million b/d in 1990 only 6.9 bcfd are estimated.

The Energy Program, meanwhile, noted that in the past as much as 12% of total gas production has been flared. The program's goal was to reduce this figure to 3% by 1990. It would be to Mexico's advantage to discover some new domestic application, export agreement, or liquefaction technology for her natural gas. Even at 4.3 bcfd, which the program projects for 1985, this is 2 bcfd above projected exports and domestic demand.

López Portillo's Recovery Program

OPEC pushed Mexico into becoming a significant producer in the international oil market—a market in which Mexico had once before, in the 1920s, held a position comparable to Saudi Arabia's in the 1970s. To be sure, OPEC forced Mexico back into the international oil business unintentionally. The 1973 oil embargo resulted in a historic break in economic rhythms between the U.S. and Mexico.

Prior to 1973, Mexico's inflation was in the 4–5% range, not so far from that in the U.S. The peso was generally thought to be worth U.S. 8¢, and the balance of payments deficit never reached even one billion dollars. But the inflation imported as a result of the oil price shock pushed Mexico's inflation in 1973 to 12.4%, a rate that doubled the following year.

The resulting gaps between the inflation rates in Mexico and those in the U.S. put pressure on the peso, since goods and services were increasingly cheaper in Houston than in Mexico City. This process resulted in the traumatic devaluation of the peso in September 1976 from 12.5 to the dollar to 20 by year-end. The timing of the devaluation corresponded with the inauguration of a new president whose orientation, unlike that of his predecessor, was probusiness. With 90% of private companies technically bankrupt and a loss of domestic purchasing power to capital flight estimated in the $2 billion–$4 billion range, the country had no choice but to look to its long-neglected hydrocarbon reserves as a source of psychological and economic recovery.

The north star of President López Portillo's recovery program was a schedule for rapid hydrocarbon exploration and production. Under the industrial leadership of Díaz Serrano, Pemex was an overachiever in these two areas. Regarding proved hydrocarbon reserves, which were 11 million bbl in 1976, there is reason to believe that they might be increased to more than 70 billion by the end of the López Portillo term. Annual average crude production, which was 877,000 b/d in 1976, meanwhile rose 1.87 million b/d during the next 6 years, to 2.75 million b/d in 1982.

The second cardinal direction was the government's crude export strategy. This strategy had three components: price, market diversification, and contract bartering. Pricing strategy was to charge the maximum rate allowed by the highest-priced OPEC member. The decision to diversify oil export markets meant that instead of selling nearly 90% of Mexico's crude to the U.S., which had been the pattern as of 1976, Mexico would sell to other country buyers and reduce the U.S. share to not more than 50%.

In 1976 Mexico was exporting crude to five countries; in 1981 to more than 20. The U.S. share of exports was down to 50%, sufficient evidence of the success of this program objective. Contract bartering was a matter of trading the crude contract for diverse benefits to Mexico that the prospective buyer might provide: technology, foreign investment, financing, market access.

The third direction the López Portillo government followed was domestic market controls. These controls directly or indirectly involved subsidized energy prices. Gasoline, diesel, natural gas, and electricity were sold at substantial discounts compared to world prices. According to one estimate prepared by

Jaime Corredor, an analyst at the Office of the President of Mexico, Pemex's 1979 sales, had they been at world prices, would have brought in $22.986 billion, not $7.286 billion. The difference signifies the aggregate dollar value of the domestic subsidies. Nonoil producers in Mexico were offered additional discounts on industrial energy prices if their business objectives met government industrialization and export development plans.

The fourth cardinal direction was not explicitly announced in the intellectual structure and tone of the López Portillo government: the movement toward a centrally planned and managed economy. This element, implicit in the country's economy since Spanish colonial days, had been revived, strengthened, and defined in the economic precepts of Mexico's Constitution of 1917. But with a few exceptions, two being the expropriation of foreign oil interests in 1938 and the repeated tinkerings with the structure of peasant agriculture, these precepts have seldom applied outside the classroom. Nor did the constitution require the federal government to take economic initiatives in the name of development.

What was new under presidents Echeverría and López Portillo was the federal government's claim to recognition as the economy's principal and most able entrepreneur. Under Echeverría, anti-imperialistic legislation was put into force restricting foreign capital. Under López Portillo, urban, industrial, and global plans were issued goals that were to coordinate and guide the economic development of the country in general and rationalize the development of the state oil industry in particular.

Initial results of the López Portillo program were startling to skeptics in Mexico and abroad: by the end of 1977 Mexico had sold nearly $1 billion of crude, quite a jump from the $420 million worth sold in 1976.

The Energy Program reflected the desire of the López Portillo administration to rationalize and coordinate the role of the petroleum sector within the optimistic macroeconomic framework that had been set forth for the 1980s. In this framework it was recognized that hydrocarbon exports would for some time be the most important source of foreign exchange earnings. But it was also recognized that these earnings could have disruptive effects on the total economy unless long-term planning and coordination efforts were successfully carried out.

The overall blueprint for Mexico called for hydrocarbon exports to be used as the principal instrument of foreign economic policy. Hydrocarbons, in their crude and refined forms, would be sold in international markets on a selective basis. Buyers would be those countries or companies who offered some measure of reciprocity in the areas of new technology, financing, markets for Mexico's nonpetroleum manufactured goods, or investments.

The Energy Program emphasized Mexico's determination to develop an adequate capital goods manufacturing capability for the petroleum and petrochemical sectors. Toward this end, the National Industrialization Plan offered incentives for domestic manufacturers of capital goods. Mexico, in short, wanted her hydrocarbon export sales to pay for the industrialization and technological upgrading needed to make Mexico, by the end of the century, a self-sustaining industrial economy with a diversified energy base.

These export sales would also, although indirectly, create 400,000–800,000 jobs per year and contribute to regional, industrial, and agricultural efforts. The cumulative effect of the optimistic projections appeared in Mexico's Global Development Plan of 1980 as a target of 8% real growth annually in GDP for the decade.

Following OPEC pricing

By September 1981, Mexican crude production at 2.48 million b/d was about 28% higher than 1980's average figure and 153% higher than 1977's average. But in the same month OPEC's total production, at 20.4 million b/d (excluding NGL), was 40% below the postembargo peak of output of 34.1 million b/d reached in December 1976.

This was at a time when industrialized countries had broken a historic pattern: as of 1980, increases in Gross Domestic Product were not necessarily tied to increased primary energy consumption. In the European Economic Community there was a decrease in gross inland energy use by all countries (averaging 4.5%), while all but two reported gains in GDP.

In Mexico, meanwhile, a policy of riding the crest of the OPEC price wave had been followed with considerable success by the government of López Portillo. But in June 1981 at least one man saw that OPEC's wave was about to break. The state oil company's director general telexed crude buyers that Pemex's heavy crude price would drop $4/bbl. Since Pemex was selling crude on a 50/50 mix of light and heavy grades, this price reduction implied a $2/bbl drop in Mexico's average export price.

The June price reduction implied a downward adjustment of projected export earnings of $5 billion, or 25% of the $20 billion expected for 1981. This projected loss of foreign exchange earnings occurred when Pemex's domestic price subsidies would amount to 300 billion pesos ($12.2 billion) annually. Since then, Mexico has accepted price reduction as a fact of life. Mexico promptly followed OPEC's price reduction of early 1983, the government bravely commenting that the implied loss of foreign exchange had been anticipated in its budget.

Effects of the 1981–1982 oil glut

Saudi Arabia, by maintaining production in the 9–10-million b/d range, forced other members of OPEC and shadow OPEC countries, such as Mexico and the U.K., to lower crude prices. In Mexico matters were particularly sensitive because of the presidential-successionist fever that had gripped the country. Mexico took its most successful oil industry manager (once believed to be a presidential candidate himself) and made him personally take the blame for the drop in Mexico's oil prices. In short, in Mexico as in other oil-exporting countries, Saudi Arabia excepted, there was much wringing of hands and gnashing of teeth. Budgets were cut back, capital spending was put on ice, and emergency measures were brought out for increasing government cash flow and decreasing the foreign trade deficit.

In Mexico this cash flow emergency was handled in several ways: cutting government spending, tightening import controls, continuing the acceleration of the devaluation of the peso against the dollar, and slashing into domestic energy price subsidies.

Not that the idea of raising domestic prices for motor fuels was such a new one in countries like Venezuela and Mexico. The concept—which never became much more than an abstract one—had been discussed for years, but no government in power could bring on itself the political suffering that domestic price increases would inevitably cause. The López Portillo government courageously took this step: regular gasoline (accounting for slightly less than 90% of gasoline consumption) was raised 114%, unleaded gasoline 43%, and diesel 150%. President de la Madrid's administration has continued in the same vein. On 7 April 1983, diesel was raised to 14 pesos/liter from 10, regular gasoline to 24 pesos from 20, and unleaded to 35 pesos from 30. Based on 1982 sales, these increases will generate $2 million additional income/day (300 million pesos).

Secondary petrochemicals

While the blossoming of the Pemex basic petrochemical industry national policy took place during 1976–1981, the private sector hardly stood placidly on the sidelines. It was busy producing its own giant industry in the secondary petrochemical field. Some of the early pioneers in the private industry have sunk from view due to mergers. Monsanto's Mexican subsidiary merged with Resistol, S.A., in 1971, and B.F. Goodrich's Mexican unit is now part of Cydsa of Monterrey. Other companies in the original vanguard have continued to expand. Dupont has hundreds of millions of dollars invested in Química Flor, Pigmentos y Productos Químicas, Nylon de Mexico, and others. Total investment in Mexico's secondary petrochemical operations was $696 million in 1975 and grew to $900 million in 1977. By 1982, that figure was more than $1 billion.

In 1981, more than 350 private firms produced petrochemicals in Mexico. Those wishing to buy basic products for secondary manufacturing had to apply to the Secretariat of Patrimony and Industrial Development. Most of the permits issued in the last half of the 1976–1981 period went to companies that established new plants in industrial development zones created by the National Industrial Development Plan, primarily in the states of Veracruz, Tamaulipas, and Michoacan. Companies in those zones have been receiving a 30% discount on the cost of energy and feedstock raw materials. The value of these subsidies was reduced when in 1982 Pemex raised prices an average of 47% on 31 basic petrochemicals, including 61.5% on acrylonitrile, 174% on sulfur, and 60% on methanol.

National Security

Of the three principles invoked in this early history of U.S.-Mexican energy relations—linkage, good neighborliness, and *ex gratia*—which proved to be the most effective? The principle of linkage was offered by the American government in the form of a *quid pro quo* argument: American diplomatic recognition in exchange for extraterritorial privileges, namely economic and juridical immunity from Article 27.

Mexicans, who were intent on eliminating islands of foreign interest in the national polity, ceremoniously rejected this argument in one full-dress pronouncement after another: The nation's laws cannot be violated for the sake of the personal, economic welfare of those individuals who, by freely choosing to come to Mexico, thereby signaled their willingness to abide by its laws.

"Yes, but—" was the American reply. The history of the principle of linkage in U.S.-Mexican relations has been that of inconclusive rhetoric productive of little beyond self-delusion and suspicion.

The principle of good neighborliness was advertised by the Roosevelt administration as one way to get around the juridical mindset and isolationism of Spanish-American states. It was the unique relationship that furnished the basis of extra and paralegal understandings. During the 1930s the argument sold better in the U.S. than in Latin America. The analogous Mexican argument, that historically northern Mexico and the (American) Southwest constitute a single economic region with a shared culture, language, religion, and labor pool, underwrites Mexico's border industrialization programs as well as its stance on the illegal immigration issue.

The argument is historically, culturally, and economically valid; but now that the shoe is on the other foot, the American government finds that the fit is not quite right. The future market on this principle in U.S.-Mexican economic relations may be forecast as that of a low-growth industry: neither the U.S. nor Mexico is willing to be as neighborly on critical issues as the other might wish.

The third principle, *ex gratia,* offers promise and room for imaginative negotiations. The unhappy controversy over a price mechanism for Mexican natural gas exports to the U.S. is a case in point. Because energy prices in both countries are political (and ideological) tokens as well as economic realities, no single, narrowly defined economic argument will fully meet the quite different scenario requirements of the two governments. Yet some agreement on a price mechanism for natural gas is unavoidable; for the gas, if it is not to be flared, must be sold to the Americans in order to free the petroleum-led recovery program in Mexico.

The Mexican scenario requirement meanwhile is led by the revolutionary mandate not to knuckle under to American demands. To maintain domestic credibility on this point, the Mexican government officially must hold to a line of hard, nonconvergent rhetoric. American hopes for a binational consensus model of either discussion or academic research are certainly going to remain unfulfilled.

If one keeps in mind that the word Spanish for "please" is *por favor* (as a favor) and that in the oligarchicality of the Mexican mind the favor relationship is often more binding than a force-of-law relationship, one might suppose that a comprehensive energy arrangement based on the force of "please" might receive more voluntary enforcement than agreements based on the force of treaty or executive agreement. Of course, in a sense such an agreement already exists. The U.S. government surely recognizes that the Mexican government's primary concern since the Revolution has been domestic self-legitimation. This concern becomes even more acute in conditions of depreciating policy credibility. No one can doubt the existence of "please don't" silently spoken by Mexico to the U.S. with regard to open discussion of political, economic, or military intervention. In return for American silence, Mexico remains to a large degree flexible with regard to its crude export policies toward the U.S.

The growing political and economic instability in Central America is Mexico's principal concern in the areas of national and energy security. Mexican armed forces have been deployed to the southern border areas, ostensibly to keep out unwanted refugees but mainly to ensure firm military control over the region. On January 24, 1982, an army general was nominated as governor of the oil-rich southern state of Chiapas. The Mexican government, meanwhile, has gone to great lengths to downplay the growing threat to national security.

The starting point for an understanding of Mexican national security concerns is Cuba. Mexico's exclusion of Cuba from the Pact of San Jose, which grants oil on preferential terms to Central American and Caribbean nations, and the Franco-Mexican initiative bestowing political status to Salvadoran insurgents, raises a complex set of questions about Mexico's oil and national security policies.

A logical starting place for a discussion of these questions is the proposal for a World Energy Plan that López Portillo presented to the United Nations on September 27, 1979. The basic principle put forth in the proposal was that some basis of equity should govern the availability of oil so that smaller nations are not unjustly penalized on account of accidents of geology. All nations, in short, had a political right to the planet's hydrocarbon reserves, and some mechanism beyond that of the marketplace needed to be found to guarantee that right. Within this frame of reference, Mexico and Venezuela in 1980 signed the Pact of San Jose, guaranteeing 160,000 bo/d to their poorer neighbors.

Also in 1980 Mexico published its own national Energy Program, which covered both domestic matters and export policies. The three main points set forth in the discussion of domestic matters concerned the need to curb the growth of domestic consumption, the need to increase energy efficiency in industry and everyday life, and the need to raise domestic energy prices gradually to 70% of world prices. As for export policies, the program set 1.5 million b/d as the annual export ceiling for the 1980s, with no nation receiving more than 50% of Mexico's crude oil exports and no nation, Central American and Caribbean countries excepted, depending on Mexico as a source of supply for more than 20% of its oil imports.

A strict interpretation of these two documents points, sooner or later, to Cuba. One might say, from the perspective of these documents, that the Soviet source of Cuban oil supplies is both a geopolitical anomaly and an instance of gross industrial inefficiency. These documents do not, however, suggest specific steps that might be taken with reference to Cuba nor on what timetable anything might be undertaken.

Some events seem to speak for themselves while others do not. A technical assistance agreement between Mexico and Cuba has been in effect since Fidel Castro's visit to Mexico in 1979. In April 1981 the Mexico City press reported that Mexican oil technicians had discovered an undisclosed but significant amount of oil in a 10,000-ft well located about 14 miles from Havana. This report was not confirmed by Pemex, which cryptically said that all details would have to come from the Cuban government.

The 17 April 1981 *Wall Street Journal* reported that, in line with Mexico's objective of channeling Cuba's interest in the Caribbean into "positive, nonmilitary pursuits," an offer had been made to explore jointly for oil and natural gas in Cuba and to build a refinery in Havana. No comments have been made about Mexico's supplying crude oil to Cuba.

Cuba meanwhile imports most of its oil from the Soviet Union under an agreement renegotiated every 5 years on terms in which the price of oil is linked to the price of sugar. Projecting from 1970–1978 data on Soviet oil exports to Cuba, the 1983 figure is likely to be 250,000 b/d.

Mexico's geopolitical concerns

What is the connection, if any, between Mexico's overtures to the Soviet Union and Cuba and Mexico's legitimate national security concerns in Central America and the Caribbean? The key to this question is Mexico's perception of Cuba.

For Mexico, Cuba is a typical Latin American state: solid *caciquismo* (iron-fist politics), long on Third World anti-American rhetoric, short on foreign exchange earnings. The difference with Cuba is mainly the way it is officially perceived by the U.S. and the Soviet Union.

But Mexico has not maintained diplomatic relations with Cuba all of these years out of any great love for its *cacique* nor out of any admiration of Fidel Castro's long-winded pronouncements. Until recently, Mexico's motivation has been cosmetic. During the early 1960s the U.S. broke diplomatic relations with Cuba and encouraged everyone else to do the same, which was reason enough for Mexico to maintain them. Recently, however, Mexico is finding ideological, economic, and military self-interest in deepening and perfecting its relations with Cuba.

Mexico first wants to protect its southern and eastern oil fields. To carry out this objective, political stability in Mexico's neighboring countries must be guaranteed. For this reason, Mexico and France recognized the insurgents in El Salvador as a "representative political force." Commenting on the Franco-Mexican declaration of August 28, 1981, Mexico's former ambassador to France, Carlos Fuentes, commented that "neither France nor Mexico want a Sovietized regime in El Salvador but neither do they want a U.S. protectorate that is fated to suffer relentless violence."

Should the rebels win without outside intervention, Mexico can anticipate policies favorable to Mexico's needs. On the other hand, should rebel leaders need support or asylum, Mexico—not Cuba—will be at the head of the list. By taking a leadership position in the area of political rights of opposition parties, Mexico upstages Cuba.

What Mexico does not want in its natural sphere of influence is another Angola in which Soviet arms and Cuban soldiers are employed in the defense of "representative government." Guatemalans harbor bitter feelings against Mexico over its support of the independence of Belize against the historic claims of Guatemala. Belize, meanwhile, without British military support is vulnerable to guerrilla warfare operations. Inviting British marines to defend the sovereignty of Belize would be complicated, constituting at the very least a violation of the Monroe Doctrine. Mexico could send additional troops to its borders but, in the event of escalation, the military situation would resemble the ill-fated defense of the Vietnamese border against Viet Cong incursions from Cambodia.

The Yucatan peninsula is not completely immune to nationalistic aspirations of its own. In 1847 during the American invasion of northern and central Mexico, Yucatán seceded from Mexico and asked for admission to U.S. statehood. On economic grounds, Yucatán, with its vast offshore oil reserves, is more viable than Belize as a state, and Canadian-style separatist movements could emerge in the future.

While there are already reasons enough to support political unrest anywhere in Central America and the Caribbean, military hardware and leadership have been lacking. This is where Cuba comes in or, unless contained, could come in either on its own initiative or as part of a Soviet initiative. Mexico's basic policy is to thwart such initiatives and to remove the conditions in which they might arise.

Mexican-Cuban trade possibilities

Although data on the past few years are not available, Mexican-Cuban trade has not assumed any great proportions. In 1977 total two-way trade was $37 million, 92% being Mexican exports to Cuba. In 1978 total trade dropped to $27 million, and in 1979 to $11 million, of which Cuba's exports to Mexico constituted $4 million or 36%.

Looking at prospects for trade between Mexico and Cuba, sugar is a Cuban export for which a market exists in Mexico, whose own sugar production in 1981–1982 was at least 150,000 tons short of an early 1981 projection. Mexico's total 1981 requirement for imported sugar was about 600,000 tons. What remains to be seen is under what concept, and in what amount, Mexico might contemplate the sale or exchange of crude oil and petroleum products with Cuba.

The Alaskan connection

Recent discussion in some sectors of the U.S. and Japanese business press has focused on the prospect of a revision of U.S. legislation prohibiting the export of U.S. sourced oil. U.S. House of Representatives bill H.R. 4346's argument is that U.S.-Japan trade would be enhanced if Alaskan crude oil were swapped with Japan in exchange for an equal amount of crude owned by Japan. Here the term "owned" refers to contracts that Japan has with third parties such as Mexico. It is proposed that Alaskan oil be exported to Japan on a trade basis for the equivalent oil for which the Japanese have contracts in Mexico and the Middle East.

Although this proposal is not likely to be acted upon favorably, owing to pressure groups in the U.S., its logical and commercial implications should not be overlooked.

With regard to that part of the proposal treating a possible swap of Alaskan oil for Japan contracted oil from Mexico, there is no sign that Mexico would agree to such a deal on a strict barrel-for-barrel basis. On the contrary, Mexico has always opposed the trader's approach to international oil sales. She has wanted to know that an oil shipment to a given buyer was going to be refined and marketed by that buyer, not simply traded on the spot market.

A second matter is price and terms of trade. Mexico typically charges a premium for its oil compared to Middle Eastern equivalent grades. As late as the fourth quarter 1982, Mexico's oil was slightly overpriced in the marketplace, despite two reductions. Japan's contract is 160,000 b/d with an agreement in principle to increase it to 300,000 b/d as circumstances permit.

Several billing arrangements are possible. Japan might not agree to pay Pemex for oil the U.S. ends up lifting. Where Mexican oil is involved, the U.S. could be asked to pay Pemex's invoices to Japan, since invoicing the U.S. for Japan's volume would likely exceed the 750,000 b/d country limits set in Mexico's Energy Program for the U.S.

The savings U.S. and Japanese consumers theoretically gain by reduced transportation costs could be offset by an additional premium required by Mexico to go through with the deal in the first place, a kind of service fee or excise tax on each barrel of crude oil traded. Such a surcharge could erase the potential savings in transportation costs entirely.

But supposing that an agreement were reached and Japanese-contracted oil were being shipped from Mexico to Houston in exchange for Alaskan oil being shipped to Japan, a new geopolitical principle would be introduced by the U.S. By analogy, Mexico and the Soviet Union could come to an agreement in which the USSR would fulfill Mexico's contracts in Europe, including 20,000 b/d to England, 50,000 b/d to Sweden, 100,000 b/d to France, 220,000 b/d to Spain, and 45,000 b/d to Israel in exchange for Mexico's serving the Soviet Union's contractual obligations in Cuba and Brazil.

In the past, Mexico held discussions with the Soviet Union on an oil-swap deal, but the Russians refused to accede to Mexican demands that, in addition to a barrel-for-barrel swap, Mexico be paid a premium equal to a share of the savings in transportation costs.

In such a system the same issues of price and volume parity arise as in the U.S.-Mexico-Japan swap. An additional ingredient is sugar. Mexico might be willing to buy part of the Soviet-contracted sugar harvest in exchange for a Soviet delivery in Europe of the equivalent barrels of oil. (Mexico's European contracts exceed the Soviet Union's Latin American contracts by at least 100,000 b/d).

Such agreements might be attractive to Mexico for several reasons, the most important being the additional leverage that Mexico would gain over Cuba. A second matter is style: Soviet and Mexican economic planners love arbitrary, nonmarket prices for goods and services. Mexico's discovery in 1981 of the law of supply and demand in the international oil market was not a pleasant experience. Mexico's industrialization plans had assumed a constant international demand for crude with rising prices and had not built in the required flexibility to deal with changing market conditions.

But the U.S. might not smile on a Soviet-Mexican oil deal, even if it were to achieve cost savings in transportation and even if Cuba were, by this indirect mechanism, somewhat taken out of the Soviet sphere of influence. In the U.S.

view of things, Mexico's oil exports should go, if not to the U.S., at least to U.S. allies and neutral powers. A Mexican-Soviet oil hookup might be an unacceptably high price to pay for Cuban containment.

Fair enough, Mexicans might respond. What price are you willing to pay to co-opt Cuban economic leaders? Include Cuba in the mini-Marshall plan for the Caribbean? Reestablish diplomatic and commercial relations? Withdraw the Marines from Guantanamo Bay? Take your choice, but the day has long since passed wherein Cuba, by being politically and commercially isolated, can be left free to upset the political and military equilibrium of Central American and Caribbean states. Containment is not the Mexican way of doing business, either at home or abroad. Mexico's way is to bring the opposition into the system. It doesn't matter how you do it; cooperate with us in bringing Cuba back into the system, or we will do it by ourselves in ways you may not like.

The Díaz Serrano connection

On September 24, 1981, the Mexican press reported that Pemex's ex-director general, Jorge Díaz Serrano, was officially being proposed as ambassador to the Soviet Union. About the same time, Díaz Serrano, in a speech in Tijuana, said that he would not be returning to industry but henceforth would be a political servant of the state. If Díaz Serrano's mission in Moscow was to draw on his experience at Pemex, then he could have served as special consultant on industrial, technological, and commercial aspects of Soviet-Mexican-Cuban oil relations. No official indication of an oil mission has come out, and informed Mexicans generally say that sending him to Moscow was more in the nature of a political exile, in part to keep him out of presidential politics during 1981–1982 and in part to wash out some of his public image as a businessman whose U.S. connections are too thick to suit Mexican political tastes.

Mexico gambled a good deal of its economic and political well-being on the assumption that crude oil prices would continue to rise at rates observed during 1974–1980. Although there is no direct correlation between political and military instability in Central America and leveling crude oil prices, the coincidence of the two developments puts an extremely difficult financial, political, and military burden on Mexico.

Japan's $1.50/bbl surcharge: insurance premium or foreign aid?

Mexico has the late Shah of Iran to thank for many of its oil supply contracts with new clients. These country and corporate clients sought to diversify their sources of imported crude. Fear of a disruptive blow-up in the Middle East was their principal motivation. By mid-year 1983 this fear had largely subsided, even in the minds of country clients like Japan for whom the issue of security of supply is a much larger one than it is in other countries (Japan imports nearly all of its oil).

Japan had cultivated commercial relations with Mexico assiduously in the years since 1979, and while negotiations were long in materializing, for a time there was talk that Japan might become a 300,000 b/d client—lifting 20%, in that case, of all Mexican crude exports. But in early 1983 Japan reduced its volume to 110,000 b/d, even though the contract level read 160,000 b/d.

Japan waited for Mexico to lower its export price in line with the inevitable reduction in OPEC's marker price. But Mexico's reduction of the Isthmus price from $32.50/bbl to $29/bbl was not enough, everything considered. Japan wanted a price of $27.50/bbl and asked Pemex to consider Japan's special relationship with Mexico.

It was, after all, a risk for Japan to lift from the Salina Cruz terminal at all, given that, unlike the Middle East, there was no alternative port to which Japanese tankers could be sent in the event of difficulties with price or volume. Further, because Mexican crude has a high metal content, it was ill suited for Japan's refinery system, which lacked adequate coking capability. Mexican crude therefore had to be mixed with oil from other sources, implying additional costs of storage and financing. However, given a price break of $1.50/bbl, Japan could absorb these costs and come out even.

MEXICO: That's fine, Pemex officials replied. But don't you see that Mexico is selling Isthmus crude at $29/bbl to U.S. and European clients—and it's completely out of the question to imagine that Japan could receive a price that was not given to everyone.

JAPAN: We understand your problem, but we don't see it as an insuperable one. Mexico did succeed in finding, in August of 1982, a formula to offer the U.S. Strategic Petroleum Reserve 100% light-grade crude at a substantial discount. Japan is no less a friend of Mexico than is the U.S., but Japan cannot raise her oil imports from Mexico if the present price persists—on the contrary, we may have to reduce our volumes further.

MEXICO: But, of course you'll keep in mind that the amortization of Mexico's investment in the Pacific Coast terminal, which saves Japan considerable transportation expense, depends on an adequate volume of crude exports to East Asian customers, above all, Japan.

JAPAN: Of course, we will keep that in mind, but we hope that you will find a way to give Japan some price relief, if not in crude then possibly in product.

In these discussions both Mexico and Japan walk a fine line. Japan, for its part, continues to want security of oil supply, and Mexican oil, despite its drawbacks, fits strategic guidelines. From the point of view of Japan's Mexican

Petroleum Importing Company (MEP), the oil deal between Japan and Mexico is at bottom a matter of negotiations between governments. Given Mexico's widely advertised policy of uniform export prices, there was no way to comply with Japan's request for a straight $1.50/bbl discount. However, Japan is too good a customer, all things considered (liberal financing from Japanese banks, for example), to risk stepping too hard on Japanese toes. For Japan, on balance, it makes little difference whether the $1.50/bbl, if it has to be paid, is a premium on a regional risk insurance policy or disguised foreign aid.

This fencing over terms of oil trade takes place, as the principals are well aware, in the shadow of Mexico's de facto oil production and export agreement with OPEC.

11

Prospect

Mexico began 1981 as a non-OPEC hawk in its crude oil prices and policies, but by midyear the softening of the international oil market prompted two cuts in Mexico's export price for its heavy Maya crude. Shortly following the second price cut, Díaz Serrano was removed as head of Pemex, but no credible explanation of this action was ever officially given.

Pemex average export sales reached 1.1 million b/d in 1981, up considerably from 828,000 b/d in 1980 but less than the 1.5 million b/d projected. As a result of weakened market conditions, Pemex's 1981 export revenues reached about $15 billion, up 50% from 1980 but short of an earlier $20 billion projection. In response, the Mexican government cut 4% of its planned expenditures and reimposed an import license requirement for 80% of its trade volume.

By the beginning of 1982, Mexico was experiencing serious cash flow problems. Pemex's expansion plans were gradually put on ice. A sharp squeeze on imports was put into effect, not only for the consumer but for Pemex, which found that import permits for millions of dollars of goods had been cancelled.

In the year that followed the publication of the Energy Program, three events occurred that affected the probable outcome of development programs: 1) softness in the international oil market that led to a falloff of Mexican oil exports, 2) management disputes in Mexico's oil industry between business and policy managers that led to the removal of Pemex's head in June of 1981, and 3) the renewed determination to achieve the basic program, expressed by the selection of Global Development Plan author Miguel de la Madrid as the successor to López Portillo.

The question that the new government must answer is, would Mexico's wisest course be to implement or to scrap the oil-keyed blueprints for economic development that were so carefuly worked out during the tenure of López Portillo? The analysis that follows suggests that scrapping would be the wisest course.

The Energy Plan conceptualizes the composition of domestic energy supply in 1979 and 1990 (Table 11.1). Each source is shown by reference to its relative contribution to the whole as well as to its value in barrels or equivalent barrels of crude oil.

Table 11.1
How Mexico's energy mix will change

	1979		*1990*			
	(Million b/d)					
	Share (%)	*Crude Oil Equiva-lent*	*Share (%)*	*Crude Oil Equiva-lent*	*Variance (%)*	*Annual Compound Growth Rate (%)*
Crude oil	58.0	1.044	58.2	2.561	145	8.50
Natural gas	29.5	0.531	26.9	1.184	123	7.56
Coal	4.1	0.074	8.5	0.374	405	15.87
Hydroelectric	6.2	0.112	4.7	0.207	85	5.74
Nuclear	—	—	1.5	0.066	—	—
Geothermal	0.3	0.005	0.4	0.018	260	12.35
Imports (exports)	1.9	0.034	(0.2)	(0.009)	– 126	—
Total	100.0	1.8	100.0	4.4	144	8.46

Source: *Programa de energía,* March 1980, p. 36.

The plan's forecast for 1990 takes into account the package of energy conservation measures that the plan provides. Without these measures the aggregate crude equivalent demand is estimated to be 1 million b/d higher. At 5.4 million b/d the annual increase would be 10.5% in demand, not 8.5%, and the overall increase in demand, counting from 1975, would be 200%, not 144%. It also is apparent how difficult it will be for Mexico to reduce its dependence on oil and natural gas for its energy supplies. In 1979 these two sources provided 87.5% of Mexico's energy, and in 1990 they would continue to provide 85.1%. On the other hand, coal, geothermal, and hydroelectric sources will have increased substantially over their 1979 goals.

To make money, the internal energy market has to be rationalized in its price structure. Toward this end a national crude oil equivalent accounting system should be established. The true cost of services provided either by the government or by private capital should be understandable by reference to its equivalent value in crude oil terms. In this area the intellectual framework built by the López Portillo government for understanding and rationalizing the Mexican economy is a useful beginning. The method of calculating national energy demand on a crude equivalent basis provides a uniform standard of reference by which industrial efficiencies and net present values can be compared. The logical implications of this method need to be pursued.

The government investment program outlined in the Industrial Development Plan does not indicate the full scope of public sector capital spending. Transportation and tourism investments, for example, are omitted entirely. Nor does the program budget indicate the crude oil equivalent values of the government subsidies in the economy. Such outlays should be regarded as capital expenses, not current ones.

One has the impression, moreover, that public sector investments are required in many cases because of government price controls and other restrictions on private capital. Who would care to compete with Telefonos de Mexico when the price of a local phone call in Mexico is 8¢ (20 centavos)? Who could compete with the transportation services provided by the electric-powered metro in Mexico City when the fixed price for any distance is less than 10¢ a round trip?

At the place in the Energy Program where it says that electricity requirements in 1990 will be 208 Twh vs 239 Twh, one must ask at what fraction of world prices are the 272,000 b/d crude oil equivalent (represented by 208 Twh on a BTU equivalent basis) sold at on the domestic market? The government-owned power company, in business since 1938, is believed never to have made a profit in any year. The Energy Program indicates that of the 1 million b/d of crude oil equivalent energy to be saved by 1990, 126,000 b/d will come from increases in the price of electricity, which has been heavily subsidized. A program for increasing prices by 1.5%/month has been at large since mid-1981, but political opposition has delayed its implementation.

The Industrial Development Plan says that capital spending by the government is necessary because private capital will not take the initiative. But private capital cannot take the initiative because government subsidies skew market conditions. The industrialization plan, meanwhile, wants to introduce more subsidies to create incentives for private capital—Catch 22, Third-World style.

Without question the López Portillo management team formed the most intellectually sophisticated government in Mexican history. But the acid test of leadership that its successor-government will have to pass will not be measured only by its ability to engineer the hoped-for continuity, but in its ability to manage discontinuity in the face of profoundly changed market conditions.

While the Energy Plan, the Industrial Development Plan, and the Global Plan have been criticized for their failure to deal with essential issues such as inflation rates, foreign exchange rates, and crude export pricing policy, these plans do illuminate business and policy options. During the de la Madrid rule there will be a finite number of barrels of oil and cubic feet of natural gas exported. But too much should not be expected from hydrocarbon exports, particularly if the export volume level is to be fixed.

Mexico's Role in the International Oil Market

The consensus is that world demand for oil will be satisfied easily. Yet, just as during the 1970s, there were reasons in the minds of the Iepes study team that prepared the white paper for the de la Madrid administration to believe that during the 1980s and 1990s there will be continued disagreements among the OPEC countries.

There will be those who will have large reserves and retain present prices, and there will be the rest of the OPEC countries whose levels are lower and who will seek higher price increases. The first group, headed by Saudi Arabia, will try to regularize the supply, moderate price fluctuations, and guarantee stability in the market with an eminently political and economic strategy that will permit the flow of petroleum income on which their internal economic long-term expansion plans are based. On the other hand, the limitation of reserves and the rapid growth of internal demand will cause exports to fall in countries like Algeria, Indonesia, Nigeria, Ecuador, and Gabon, further strengthening the position of the main oil producing countries.

Based on these reflections of the behavior of supply and demand, the Iepes team concluded that "neither producers nor consumers are interested, each one for different reasons, in a significant lowering of conventional hydrocarbon prices." In the future, minor price wars could occur between the major producers because of the unwillingness of some to submit to OPEC agreements to reduce crude production. However, unbridled competition for the available markets will not, the Iepes team concluded, take place.

After 1983, if OPEC unity can be maintained, price increases would be implemented only to compensate for inflation until the imbalance between supply and demand is eliminated, which was not forecasted to happen until 1985.

Substitution of petroleum by alternative sources would continue to have an unequal impact upon world demand for heavy and light crude. In view of the world refining industry's growing integration process, there were reasons (not enumerated) to believe that heavy crude will be more in demand in the medium term.

The Iepes report reiterated the López Portillo doctrine about security of supply. Because of "the uncertainty always present and the strategic importance of petroleum, even in a flooded market, those producers with better guarantees of stability and dependability of operations will be in position to maintain the advantages they have had in the past in the marketing of their products."

Mexico's reserves and commercial supplies of hydrocarbons have had different effects on the world market over the course of time. When the reserves were of less importance and demand was marginal, some buyers as well as other producers saw it as an opportunity for diversification, and others saw it as a relief from their own problems. As Mexico's importance grew, the effect was different since buyers found the opportunity to insure a part of their supplies with Mexico while sellers saw a potentially important competitor emerge.

Mexico's geographical position—coasts on both oceans, close to industrialized countries—as well as the nonreligious character of its crude (as opposed to the religious character of Middle Eastern crude) confer on it a unique competitive position. Given the parity with Middle Eastern, African, and Venezuelan prices, it should be difficult for Mexico to lose markets. However, neither Japan nor Europe, the Iepes report noted, has considered Mexico's crudes as absolute commercial alternatives, in view of the FOB price mechanisms of the Mexican ports.

Relations with exporters

Mexico has stopped being a marginal supplier of the international oil market and, given the magnitude of its hydrocarbon resources, Mexico will continue to have a front-row position in world supply. Its pricing, production, and exploration decisions have a growing impact on the world market.

This growing participation, intimately tied to long-term Mexican interests, means that Mexico will need a strategy of rationalizing the market to achieve a price level high enough to avoid transfer of petroleum revenues to the petroleum consuming countries and to confront the financial needs for short-term internal development. This should occur, the Iepes group noted, at a price level that permits neither an overdevelopment of alternative energy sources nor strengthened conservation programs, either one of which would slash the value of Mexican hydrocarbon reserves.

A satisfactory solution to the problems of relations with other petroleum-exporting countries and the need for more coordination among the international market suppliers should not only address short-term problems and the internal conditions (actual and potential internal production capacity) but also a broad horizon of long-term objectives and interests. These should include orderly transition toward other energy sources, energy self-sufficiency, and a diversification of economic dependence.

With these considerations in mind, the general focus that has traditionally guided Mexican policy toward petroleum-producing countries in general and toward OPEC in particular should, the Iepes team felt, be widened. This should be done, "not so much out of considerations of solidarity with Third World countries, but out of a consideration of the present challenges that the international energy policy of Mexico faces with respect to the world petroleum market."

Importance of the San José agreement

Concerning agreements between Mexico and other petroleum producing and consuming countries, the significance of the San Jose agreement of 1980 stands out. By means of this agreement, Mexico and Venezuela pledged their commitment to sell crude on financially favorable terms to several Central American and Caribbean countries.

Even though the agreement is, the Iepes team felt, "a gesture of unprecedented generosity, given the hardships the Mexican economy is going through, its importance is greater than any economic consideration and is to be understood in the light of the broad objectives of Mexican foreign policy." It gives credibility to the Mexican commitment to a new international economic order that favors cooperation between developing nations.

Iepes felt that one role played by the San Jose agreement was its function as a useful instrument in response to "U.S. government pressures to incorporate Mexico in its plans for the Caribbean Basin." The San Jose agreement should be considered, they stated, as solid proof of Mexico's search for mechanisms to help overcome the difficult economic and political problems of Central America. At the same time, Mexicans value this agreement for being expressed independently and for affirming the principles that Mexico has traditionally defended—political nondiscrimination in the administration of aid programs.

Strategic Petroleum Reserve

In respect to the agreements to sell petroleum to the U.S. Strategic Reserve made in August 1981 and 1982, there is no doubt that these can help Mexico emerge from its extremely difficult economic circumstances. However, these agreements, the Iepes group observed, will have an effect on the relations between Mexico and other petroleum-exporting countries. Mexico should make an effort to coordinate an understanding with these countries as a first step toward the rationalization of the market.

For this reason, Iepes recommended to the de la Madrid transition team that the government should avoid concluding contracts that undermine the bargaining position of other petroleum-producing countries and awaken doubts about the solidarity of Mexico with the principles held by OPEC countries with respect to supplying crude for the strategic reserve. The Iepes group was concerned about the hydrocarbon and mineral resources that extend across the border and across the continental shelf divisions of Mexico and the U.S. They felt that the exploitation of such deposits could provoke conflicts between the two countries, which should be anticipated.

Package deals

Concerning package negotiations, Mexico can, they thought, continue to get advantages because of the adverse internal economic conditions and the recession among industrialized countries, provided that Mexico studies the best way of linking petroleum negotiations to other aspects of bilateral exchange. Package agreements should be examined casuistically, and despite the international recession, the team felt there were opportunities that still had interesting possibilities.

Marketing ethics and pricing policies

The Iepes group, echoing national ideology, voiced the view that Mexico has been the exception in the petroleum world, in that its marketing policies have been guided by invariable criteria that are unique among petroleum-exporting countries. "Mexico treats its clients equally, having one price for all. Further, Mexico does not discriminate on political grounds."

Yet many observers would say that Mexico's reluctance to conclude a swap deal with Russia, Mexico supplying Cuba and the Soviet Union supplying Mexican contracts in Europe, is blocked by political constraints, even though much has been made of the commercial difficulties. Initial reluctance to sell oil to South Korea was also said to be on political grounds, Mexico not wanting to be seen as an indirect supporter of an authoritarian, repressive regime. Most important, unlike other countries, "Mexico has refused to let its crude oil become a spot market commodity available to speculators," say Iepes analysts.

Without some appreciation of Mexican history, such a statement borders on the irrational, as if crude oil were not a product for commercial transactions. Mexicans, since the days of the Spanish colony, have felt that the wealth of the earth belonged to society at large. In the colonial period the Spanish crown had exclusive dominion over subsoil wealth, and this idea—or, better, feeling—was put into Article 27 of Mexico's constitution of 1917. Mere traders, in short, should not be permitted to get their hands on Mexico's national patrimony.

The Iepes team believed that Mexico should preserve those qualities in its oil policies of the future. On the other hand, it stressed the need to remain competitive. In this area, mindful of Mexico's miscalculation and consequent loss of customers mid-1981, the team advised caution in determining the optimal price with regard to its impact on foreign markets, and Mexico's ability to maximize export income. Further, even though the strategy of requiring Pemex customers to buy equal amounts of Isthmus and Maya proved effective, if their respective prices were competitive, the crudes could be sold separately in order to enhance Mexico's responsiveness to the market.

The oil market, the Iepes group noted in wry understatement, "is not independent of international political considerations." Besides strictly commercial considerations, Mexico needs to align its petroleum policy and its foreign policy along clear and congruent guidelines. Mexico's participation in the international oil market should take place, they asserted, not only with commercial considerations in mind but also taking into account its well-defined principles in the area of international economic policy.

Flexible policies for Pemex's foreign trade

As a central monopoly organ for the hydrocarbon sector, Pemex is obligated first to satisfy completely national demand for refined products, residual fuels, special products, and basic petrochemicals. Even though reserves of the last years have guaranteed an adequate supply of raw materials, shortages in the industrial sector have caused Pemex to be a net importer of petrochemicals and lubricants.

The commercial policies that have been the guidelines for these imports have been the same as described for exports. The most significant have been the completion of acquisitions, processing agreements, and many direct interchanges with producers. Crude processing agreements have the advantage of using national raw material and have avoided foreign currency exchange transactions, eliminating inconvenient price fluctuations and speculations.[1] On the other hand, important advances in the diversification of imports should be continued and widened in coming years.

Concerning substitutions, conservation, and rationing, there is room for action by the head of the energy sector. Iepes contends that he should designate a specific body to address the problem because such projects require long maturity periods and difficult conciliation of interests.

Until recently, Pemex managed its foreign hydrocarbon transactions and equipment imports separately. It now has a manifest interest in taking advantage of exports, especially of crude, to the countries that supply equipment to the energy sector as a means of getting better financial and price terms. Thus, in its negotiations with these countries, Mexico should concentrate on import necessities and should use the potential exportation of crude as leverage to get better conditions. There are a number of examples, e.g., Japan, in which export negotiations have led to joint investment programs in Mexico for the creation of capital goods industries.

Refined products and petrochemicals

Concerning refined products and residual fuels, the current unattractive world prices for refining lead to the recommendation, at least in the short term, that export policies not be oriented around refined products. Mexico will not be able to count on a large excess refining capacity in the country.

The recession has brought with it a very significant decline in the consumption of petrochemical products in the industrialized developed countries. Thus, Mexico will eye closely the export of petrochemicals in order to bring a greater flow of foreign exchange or minimize its outflow. For example, the recession will decrease ammonia consumption in agriculture, warranting a close reevaluation of its exportability.

---

[1]The economics or politics of Pemex's acquisition of Gulf's minority shares in the Spanish refinery Petronor in 1979 and 1980 have never been revealed. The generally held assumption was that Pemex had its eye on the EEC market as an outlet for refined Mexican crude.

Sales strategies

Mexican sales strategies have surpassed those of strictly buy/sell operations. For some time, such strategies included a marginal volume for crude processing operations, an activity that has proven beneficial for covering domestic gasoline and petrochemical products shortages. However, in areas such as catalytic production, processing activities could lead to a high dependency on imports, which would be counterproductive for Mexico.

A real internationalization of the commercial activities of Pemex implies channeling greater economic resources to it and granting a greater degree of legal and fiscal flexibility that increases the degree of flexibility of its foreign sales operations. This would permit the company to be in a position to take advantage of the demand for diverse products, cut unused plant capacity, and offer excess products for export. That would mean broad government support because it would require rewriting financial laws—even to the extent of treating Pemex as a special case, giving it the maximum ability to react to foreign market fluctuations.

To put into practice new sales mechanisms and strategies that will satisfy these prerequisites, Iepes recommended that the following possibilities be studied:

- mechanisms to ease and simplify administrative procedures
- the creation of a foreign commerce unit charged with executing commercial policies in accordance with the guidelines, as defined by the executive of foreign hydrocarbon commerce
- the creation of enterprises with other countries or companies for the commercialization of products outside of Mexico[2]
- leasing fully integrated refineries that might be available in other countries as an alternative to crude processing agreements
- creation of a Pemex affiliate in some convenient location in the Caribbean

Natural gas

Perhaps the most important determining factors affecting the volume of natural gas exported during the 1980s are the advances in converting the industrial plant to natural gas, the level of crude oil production reached, and the reduction of the large losses in the petroleum fields of the Campeche Sound.

In this respect, changes in the structure of domestic prices for natural gas and residual fuels introduced in mid-1982 have as their object increasing, for the medium term, the volume of gas available for export and reducing the export sales of residual fuels whose sales price is less, in calorific terms, than natural gas.

---

[2]In the second half of López Portillo's regime, there were frequent rumors that Pemex would conclude some kind of a marketing-oriented joint venture in the U.S.

Moreover, during the 1980s Mexico will find itself in an increasingly competitive position with Canadian gas producers with regard to selling excess gas on the U.S. market and the possibility of maintaining current prices in real terms. Keeping Mexico's and Canada's interest the same as during the 1970s, concerning pricing policies and exports to the U.S., will depend on the contracts and eventual agreements that can be reached.

Export planning

Finally, in a tight world petroleum market it was possible, but not always advisable, for Mexico to hold down operations in foreign commercial areas in favor of internal market needs and easier productive areas. As of 1982, the complementary and balancing role that export operations had in the past cannot remain as a realistic role. The planning for refineries and petrochemical processes will hinge more on the needs and opportunities of the foreign market in order to guarantee as much as possible continuity and the highest foreign exchange inflow.[3]

De la Madrid's National Development Plan

Twenty-four hours before its own deadline of 30 May 1983, the de la Madrid government issued its global outline for national development for the six-year presidential term. The plan said, in essence, that within a couple of years Mexico could look forward to strong economic growth, between 5–6% a year—if. The principal if's were items outside the control of the Mexican state. If, during the 1983–1988 period, the annual growth rate in the industrialized countries occurs in the area of 1.5–3.0%, if inflation stays between 5–10% (after 1985), and if oil prices stabilize at present levels.

The qualitative approach

The thrust of the analysis was self-consciously qualitative. Where the econometric style of reasoning of the López Portillo planners had been to contrast how things would go in Mexico if left to their own devices with how things would go if their various plans were implemented, the de la Madrid team believed that it was better to keep quiet about such matters. "Mexico suffers from two things," a Mexican senator said during López Portillo's time, "the demagoguery of optimism and the demagoguery of pessimism." In refusing to give more than a single table of economic assumptions (forecasting macroeconomic statistics for the industrialized world) in more than 400 pages of text, de la Madrid planners were acting in the belief that the econometric exercises of their predecessors had been a form of demagoguery.

[3]This conclusion, that investments at home in the industrial plant must be evaluated with one eye on the potential export markets, is the inverse of traditional Mexican doctrine regarding investments in refineries. Traditional doctrine, emphasizing the need to satisfy domestic demand, had relegated analysis of export markets to a back-seat consideration.

In their diagnosis of the health and productivity of the energy sector, they found much to be desired. The patient was sick.

Ailments of the energy sector

The energy sector was suffering from vitamin deficiencies, digestive problems, tunnel vision and a vague sense of mental and moral ill-being.

Deficiencies. Because of a preoccupation with achieving quantitative goals, because of a penchant for centralization, and because of a greed to achieve as much as possible in the shortest time, energy sector management had not foreseen the profound effect of its operations, much of it negative, on rural economies and environments. Nor, in its haste, had it considered the negative impact of so many imports on the economy: When will Mexican firms produce these goods and services if state-owned energy companies pass their requirements to deep receivers in procurement offices in Houston?

Because of a policy of keeping energy prices below levels of inflation, energy became ever cheaper during López Portillo's period—despite price increases from time to time. The inevitable result was that certain resources and products—natural gas and gasoline, for example—were wasted and over-consumed. In the case of gasoline, the matter was made worse because of the refinery designs had called for crude oil in a ratio of 75% light to 25% heavy—while the country's physical resources clearly favored heavier crudes.

The result was that some products were sold below cost. This meant that the energy sector, never able to generate enough profit to pay for its own expansion, had to seek financing in foreign capital markets.

Inefficiencies. The excessively rapid growth of the sector meant that supply and demand of goods and services got out of balance, especially in rural areas. Often there was inadequate storage for crude oil and refined products. Some oil-producing fields were overworked, and the flaring of natural gas was another case of industrial inefficiency.

The electrical sector had its share of problems. Hydrocarbon energy, which represented 90% of total national supply, has progressively replaced hydroelectric sources. Fuel oil runs 65% of the nation's thermoelectric plants, but inadequate purification of the fuel has resulted in environmental contamination, higher maintenance costs, and a reduced expected life of the plants. Making matters worse, many parts and supplies were imported.

Difficulties. Although Mexico's resurgence as an oil exporter occurred under the favorable conditions achieved by OPEC, by the end of the López Portillo period Mexico faced downward pressures on oil prices. This meant that Mexico and other oil exporters were under the necessity to strengthen "mechanisms of coordination" to defend efficiently the "just value" of their products in the international market.

What to do?

With this very general diagnosis in hand, de la Madrid proposes an energy program for Mexico that accomplishes development of broad goals. To implement these goals, the government issued "strategy guidelines" that are developed under the heading of "General Lines of Action." Each of these latter, in turn, has its own shopping list of program ideas. The broad goals may be summarized as a desire to 1) increase the management efficiency of the energy sector, 2) rationalize the supply and demand of energy products, 3) emphasize the sector's role as a market for domestic suppliers of goods and services, and 4) emphasize the sector's potential role as an agent of regional development.

The idea of improving the management efficiency of the energy sector is the most intriguing of the lot, given that the government clearly means that there should be greater efficiencies, not only in Pemex, but at the "intersectorial" (cabinet) level. Unfortunately, accenting the qualitative side of life tends to turn a planning document into a nest of MBA clichés. For example, the first of the "lines of action" to consolidate and reorient the productive structure of the energy sector is to create a single authority for the sector that would "guarantee the internal congruence of the sector, permit the administrative restructuring of the sector's public companies to take place, and assure the sector's activities correspond to those of other sectors." The first-listed program concept designed to accomplish this worthy aim is the establishment of "mechanisms of intersectorial coordination and systems of evaluation and control to assure that the entities of the sector adjust themselves to the guidelines and strategies for national development as well as to those which derive from the requirements of government expenditures."

Evidently, what this all means is the real (i.e., quantitative) development agenda of the de la Madrid government is not going to be made public, at least not right away. There are good reasons for being shy in this area, the main reason being that the feasibility of major economic goals depends on too many people who don't travel with Mexican passports—the staff of the IMF, for example. Why create popular expectations if you're not sure you have the money to pay for the goals being carried out?

Energy sector policy in Mexico is formulated on two levels, skin and heart. But the skin level of policy is more than a mask covering real intentions. In this area Mexico shares many qualities with other developing nations.

12

Mexico as an LDC

Mexico's experience can be usefully compared with those of other countries. Some *Wall Street Journal* readers objected to the comparison of Mexico and Iran,[1] but it would be unwise to treat Mexican business ways as essentially American or as *sui generis* in character.

Mexico as a Great-Power Neighbor

The governmental institutions of great-power neighbors tend to either imitate or rebel against those of the great powers with the strongest influence over their national destiny. In the first case, it is imitation only as protective coloration. In the second case, the reaction is apt to be a caricature of the original model.

In either case, imitation or rebellion (one thinks of Cuba and Albania), the trait that gives it away is the monopoly on national leadership that is claimed by the central government. It follows that independent spokesmen from the private sector are not allowed to speak openly to the national or international press. This curious silence from the private sector characterizes great-power neighbors, particularly less-developed countries (LDCs).

One of the distinguishing features of LDCs is the obsession with consolidating government power. No institution, and certainly not the press, is immune from government pressure to adhere to a monolithic interpretation of the government's preeminent role in political, intellectual and economic affairs. This means that the English-language press (printed and electronic) in these countries tends to be simply a sop for tourists, one that is devoid of substantive

---

[1]*Wall Street Journal*, 9 September 1981.

intellectual content. When an issue of a foreign periodical carries an article critical of the person or family of the LDC ruler, copies of that issue are simply not circulated inside the country.

The political institutions of great-power neighbors are like chameleons that, for self-defense, take on the color of their surroundings. Since the U.S.-Mexican War of 1847, Mexico has successfully avoided further military encroachment on national territory by a remarkable facsimile of the outer forms of U.S. political institutions. This feigned political kinship has appeased Washington's needs for a satisfactory buffer state.[2]

The destinies of great powers and their neighbors are inevitably linked by bonds of trade, investment, and blood. In expansive movements government and academic spokesmen of the Great Power's institutions are apt to speak of the interdependence of the two countries, but this notion, although based in fact, is politically and ideologically unacceptable for the nation at the other end. Neighbors of this Great Power usually ignore the rhetoric of interdependence and in its place substitute an ideology of autonomous development supported by appropriate bilateral understandings and treaties.

In their symbolic relationships the matter of their blood ties is often politicized. In the Great Power, the presence, legal and paralegal, of tourists, workers, and naturalized citizens from neighboring countries takes on economic and political significance where large numbers of people, dollars, or votes are involved. Occasionally, there are instances of ethnic politics, as where a descendent of naturalized citizens is appointed ambassador to his ancestral nation. Great-power neighbors generally take exception to such political ploys, saying that they want an ambassador who is representative of the political, cultural, and economic mainstream of the Great Power.

Not in every case does the great power have to be geographically contiguous. South Korea is the great-power neighbor of both Japan and the U.S. and, accordingly, her governmental institutions resemble both countries—an emperor-type democracy. Ideologically, Korea is so anti-Communist that not even President Carter could, with impunity, fulfill his 1976 campaign promise to withdraw the Eighth Army. Cuba, on the other hand, rebels against the U.S. and imitates Russia—but with Latin bravado.

Mexico as an OPEC-type Political Economy

Mexico has been called the Saudia Arabia of America, but not only because Mexico is becoming a significant exporter of crude oil. Three other reasons have been cited:

---

[2]See Appendix K.

- Mexico's development planning is dependent on crude export earnings as the main source of public sector development funds.[3]
- The political structures are analogous.
- The national oil industry is a mystique.

There is a certain consistency involving several issues and patterns in the history and national psychology of LDCs. Although in recent decades, with the growth of international interdependence in matters of markets, currency, energy, and labor, LDC economic history has tended to blend with that of other nations, LDC economic patriots themselves stoutly hold to an autonomous model of development.

Until recently, issues of American business publications contained sections devoted to LDCs in which prospective corporate investors were dazzled by the spectacle of prosperity. Now articles are often pessimistic. But one thing is true: the top LDC families and their holdings are enjoying a major development boom, yet what is a boom for LDC oligopolists does not necessarily translate freely into an opportunity for foreign investors. Some skill at reading between the lines is necessary.

Tabooed Subjects in LDCs

Management programs treating LDCs generally avoid a number of topics that, if dwelled on, would exaggerate their importance to the foreign business observer. Forbidden topics about which books could be written include the following:

- The electoral system
- Government corruption and abuses
- The impotence of the legislative and judicial branches
- Lack of freedom of speech and freedom of the press
- Maldistribution of income
- Nepotism
- The oligarchical, plutocratic system
- The person of the president, king or ruler, including his immediate family
- Poverty, rural or urban
- Racism, implicit or overt

The public media in LDCs

In LDCs, the media tend to be tightly controlled by the government. This control is accomplished in several ways:

---

[3]Abel Beltran del Rio, "The Mexican Oil Syndrome: Early Symptoms, Preventive Efforts and Prognosis," 15 February 1980.

- Nationalization of the media
- Monopoly as a supplier of newsprint
- Restriction on foreign investments in the telecommunications field
- The use of paralegal riot police and other instruments of bureaucratic harassment where violations of public policy have occurred[4]

U.S. Business Communities in LDCs

Because multinational companies in LDCs are vulnerable to numerous paralegal sanctions for any breach of trust with the LDC government, their managers and representative trade associations or chambers of commerce often appear as more nationalistic than the LDC government.

In public, international managers must identify themselves and their companies with the development goals and policies of the LDC host government. This means that U.S. businesspersons and organizations in LDCs must learn to survive while operating in what one observer has called a quasi-hostage status.

A company learns that its future growth and earnings depend on its political cooperation as much as on its business acumen. Included under the category of present conduct is a requirement to promote foreign investment in the given country. This means that U.S. business organizations in LDCs, particularly where they have binational participation, tend to be restricted in the quantity and sensitivity of business and political intelligence they are free to discuss. U.S. embassies typically consider such American chambers of commerce as organs of the host government, not as the voice of U.S. industry in that country.

The Two-Tier System of Taxation and Accounting in LDCs

National, regional, and local taxes are collected in LDCs by a two-tier system of taxation and accounting. This system may be described in several ways. One way, by reference to the service or function performed, is to divide tax payments into two classes: 1) payments for governmental services or 2) payments for paragovernmental services.

Both types of payments are calculated as percentages of sales revenues, reported and estimated. Payments for paragovernmental services are sometimes made through the good offices of a coyote, a person who accepts tax payments on behalf of government officials. In seeking to do business in an LDC, a corporation may discover only later that the consultant, attorney, or business partner was simultaneously performing such services.

The functioning of this two-tier system may be seen in the following example. Having been audited, a company is found to owe $1 million in back taxes. After negotiations (the owner acknowledging that monies had been set

---

[4] *Wall Street Journal,* "Monterrey Firms Fear Government," 7 October 1982.

aside for reserve contingencies such as dealing with tax officials), government and corporate officials agree that a $250,000 fine will be paid to the public treasury and that an equal amount will be paid directly to representatives of the agency. After this latter payment is made but before the fine is paid, management receives a call from another agent who offers to cancel the fine with the understanding that his office would receive $100,000 once a clean bill of health had been issued to the company by the appropriate tax offices. In this example, assets of private agencies within the government increased $350,000 and the tax liability of the company was reduced to zero. The company owner had rightly reasoned that a reserve fund, not to be audited, would be instrumental in settling his tax liabilities.

Development History in LDCs

Historically, there are two classes of LDCs: those that became independent nations in the nineteenth century and those that became independent during or after World War II (the majority). Chronologically, Mexico belongs to the first group, but psychologically it belongs to the second. March 18 is Mexico's real Independence Day (on March 18, 1938, foreign oil interests were nationalized).

In both classes, the history of LDCs tends to be one of heroes. LDCs have developed their own style of hero psychology as part of their national culture. In some cases, the first heroes were immigrants who succeeded in implanting their language, writing system, religion, and psychology of government. In the history of other LDCs like India, the indigenous threw off the yoke of foreign usurpers. After a period of conquest, the heroes settle down to a comfortable life based on a labor-intensive exploitation on the surface, and in Mexico's case, subsurface wealth of the country. In time, the heroes tend to become institutionalized.

Although the history of LDCs tends to be one of heroes, and although the psychology of national wealth has been and continues to be based on mercantilistic theories requiring the control and exploitation of national resources, the names of the heroes are changing. In the process of becoming independent, managerial, bureaucratic nationalism becomes the new order of the day. Aristocratic nationalism, based on blood loyalty, is over and ecclesiastical nationalism, based on a supposed consensus on matters relating to the organization and control of national values and beliefs, is on its way out.[5]

Corporate hero-making in the public and private sectors has been a relatively recent development for nineteenth-century-born LDCs—one that has come to supplement the tradition of making heroes out of generals, priests, and bureaucrats. Heroes can be made out of anything that promises to fuel the LDC

[5]A long-time foreign resident of Mexico commented that the Mexican government's relationship to society is analogous to the papacy's relationship to Catholicism.

passion for images of national unity: gods (but especially goddesses), days (of battles, etc.), plans and programs (for social reform, national development, etc.) and even statistics (showing national wealth and power). It is no wonder then that hero image-makers have sometimes become heroes in their own right. Corporate advertising in LDCs is an elitist profession.

Revolutionary movements produce their own elite corps of managers, as the history of Communist bloc nations so amply illustrates. Somewhere after the first transformation that eliminates blood nationalism and before the second transformation that eliminates the legal basis of ecclesiastical authority, U.S.-LDC relations go sour, and the conspiratorial orientation of LDC xenophobia shifts.

U.S.-LDC Relations

The details of the events leading up to the souring of U.S.-LDC relations are unimportant in themselves. They are important for being cast in bronze in the popular psychology of modern-day LDC patriots. Perhaps it did not have to turn out this way, but given LDC patterns of institutionalized xenophobia and the American tradition of this-worldliness (in contrast to the supposed other-worldliness of LDC societies), someone eventually was bound to recognize that the conservative values of LDC society were in contrast and potential conflict with the entrepreneurial values of American society.

When someone on the LDC side finally recognizes that the domestic and foreign policies of the two societies are based on different principles, it is often too late. American spontaneity usurps LDC claims to several secondary markets (one thinks of Texas). LDC pride is hurt and things turn nasty.

The result is that a new international demonology arises in LDCs, one in which the U.S. government and U.S. multinational corporations emerge as central figures. In the nationalistic, anti-imperialistic mythology of LDCs, the U.S. is painted in varying shades of avarice, cruelty, and perfidy.

In speaking of LDC demonology, it is only fair to allude to its American counterpart. In the American political economy the threat of international communism is not only marketed in itself as a valuable political commodity, one useful for advancing the careers of politicians, but is also used as the marketing rationale for major government contracts—in the aerospace industry, for example. To LDCs, this American demonology mainly serves as the pretext for maintaining armed forces on a massive scale that will intimidate weaker nations. Every so often in the national media of LDCs, some allusion is made to the danger of an American military strike against the nation's natural resources. The purpose of such remarks is not to raise the level of military spending in LDCs but to flame the coals of a quiescent anti-Americanism.

In consequence, American history, culture, and life are seldom studied for their own sake in LDCs. The American political economy does, however, receive ample curricular and media attention on occasions where case studies are needed in matters relating to modern imperialism and transnational capitalism.

The frequency of such occasions varies with the frequency of corporate management crises in LDCs. When things are smooth sailing in LDC economies, interest in America is likely to be limited to means of attracting the foreign exchange of tourists and venture capitalists. But when domestic political waters get choppy, the latent demonology reemerges.

There is also a latent anti-LDC prejudice in the U.S. The colors in which the national character of LDCs is painted are sometimes less than flattering, and on certain issues the kettle of American impatience nears the boiling point. LDC nationalists are subliminally aware of this negative cultural bias but generally write it off, and rightly so, as a product of ignorance and ethnocentrism.

The psychological and cultural economics of the U.S.-LDC relationship boils down to a real problem of investment intelligence for the American corporate investor:

- He is faced with the problem of distinguishing the real from the official in terms of data, criteria, and market potential
- He has to learn to live with significant discrepancies between his numbers and those officially sanctioned
- He has to avoid stimulating competition by his very expression of interest

Mining Colony Politics in LDCs

Many LDC leaders are often economically still living in the sixteenth century: the state as a mining colony. For centuries the state has existed as a holding company or regulatory agency of the mining industry. In the current and previous centuries, LDCs have been leading exporters of what might be called "monetizing minerals," i.e., minerals that have served as the basis for comparing the value of national currencies. Current LDC development planning tends to be in line with this tradition. A national mineral in its raw and semiprocessed forms is once again scheduled to be the motor of economic revitalization. The *raison d'etre* (beyond bureaucratic self-preservation) of current LDC administration is to facilitate this process.

Once this simple fact of LDC history and economic psychology is understood, many perplexities disappear:

- The institutional hero for the LDC economy will superintend the exploitation of the currently favored national mineral. The hero will require the lion's share of the national budgeting and international publicity. (Managers and workers of the hero institution will also receive special economic favors).

- The reliability curve of economic data and criteria is roughly bell shaped, with the height achieved during the middle years of the tenure of the national holding company officers.
- The corresponding credibility gap between public, semipublic, and private information is therefore that of an inverted bell curve with the least gap occurring during these middle years.

Some mention should be made of the top management recruitment process of the state holding company. Every 4 to 6 years a new executive director is chosen, usually from among the ranks of top-level managers. The selection process is a compromise between workers, managers, and owners, but in fact little or no public information is available on the actual selection process. During the past half-century the holding company has developed a relatively effective corporate communications strategy for domestic and foreign media markets. One component of this strategy is a ritual or ceremony in which the company's workers and dependents affirm their concurrence in the selection of the new company executive. A second component calls for waves of progress reports framed in a language that suggests the new executive director is personally responsible for contributing the missing link in the chain of corporate decision-making.

Something is always said to have been missing, the same something that the new director provided upon accepting the position. LDC public relations officials are the bards of LDC society, and stories of company presidents are told everywhere. These stories tend to be the only thing known about the decision-making process. Top LDC company officials, unlike their American counterparts, do not publish their autobiographies or diaries, and therefore little is really known about inside company politics. For reasons not fully explained, LDC citizens sometimes make the new corporate officer the target of friendly, and varying degrees of unfriendly, jokes that are passed around in lieu of facts.

Attitudes toward Foreigners

Ambivalence toward foreigners in LDCs is almost a way of life. In one LDC the story goes that once upon a time a beautiful princess opened her heart to a foreign lord who had come to vandalize her country (Malinche and Cortés). This popular tale is told to school children as if to warn them of possible dangers in joint ventures with foreigners. The laws and economic development rhetoric of LDCs are also designed to reinforce an attitude of suspicion of and intolerance toward foreigners and their condescending ways.

LDC xenophobia often expresses itself as rhetorical and sometimes radical ambivalence toward foreign investors. On one hand, LDC industrial and financial oligopolists have a qualified need for the high technology and venture capital of foreign investors, particularly in economic activities of the state holding company related to natural resources. On the other hand, the national culture is committed to a development rhetoric stressing self-reliance and national independence. The result is that the deck of investment laws, regulations, and procedures is stacked against the foreign investor. To counterbalance this negative situation, a variety of government-sponsored incentives is advertised for investors, domestic and foreign.

The trust system in LDCs

In LDCs major business decisions arise from interpersonal networks (of confidence or trust, as they say), not out of formal systems of control documentation. In LDCs a proposal is acted upon not so much on the strength of the evidence and argument contained in it as on the strength of the trust relationship between the persons involved. In such a business culture interpersonal intuition, nonverbal communication, and political savvy are highly valued; in contrast, mere data is undervalued for being too impersonal. Because of the high value placed on unique trust relationships, it is natural and perhaps inevitable that the extended family system should continue to thrive in the business community.

The private, family-owned business is still the preferred model for small and large corporations. By keeping top management in the family, trust relationships are by definition guaranteed. This core of trust relationships is surrounded by less-than-trust relationships colored in varying shades of paranoia.

Because the U.S. is not "in the family," there is no basis for trust; besides, in LDCs anti-Americanism is virtually institutionalized as a vent or escape valve for frustrations produced by a domestic business environment in which there are islands of trust surrounded by oceans of paranoia. LDC paranoia vis-á-vis the U.S. is only a projection in international terms of feelings and attitudes characteristic of interpersonal, interinstitutional, and intersector (church-state-industry) relationships.

In the LDCs this trust system works, especially for the colonies of Havé's in the midst of masses of Have-Not's. For a foreigner, to insist that the political and economic system of LDCs operate responsibly by actually following impersonal systems of financial accountability is to insist that the trust quality that lubricates LDC society is no longer relevant to modern corporate life. It is to insist that dollars and records are valued more than human relationships.

And there is another problem: the government assumes that for tax purposes reported income is a fraction of ½–¾ of probable income. Businessmen fear that if they were to report real income, they nevertheless would be suspected of understating it, given the character of their political and business cultures. In LDCs political and economic inbreeding is the rule, and exogamous mating, whether in multilateral associations (such as GATT) or in joint ventures with multinational corporations, is generally frowned upon.

In some cases these tribal costumes are quite elaborate. Borrowings from the U.S., the Soviet Union, or Western Europe are common. LDCs cultivate the myth of the variant in which there is a dominant, consensus party to which the majority voluntarily belongs. On rare occasions, rulers of these consensus parties will pronounce in favor of minority party development and representation. The short-lived Hundred Flowers movement in the People's Republic of China (PRC) is an example.

Law and lawyers in LDCs

In LDCs the actual political process is generally that of a closed society. LDC governments often go to great lengths to disguise, by means of constitutional, representational, and electoral costumes, their secretive, tribal rituals.

The legal system in LDCs is therefore quite unlike that in the U.S. In a society in which only the executive branch has effective authority, the literary products of the legislative and judicial branches cannot be regarded as the LDC counterparts to U.S. laws and judicial precedents and rulings. Such literary products are the expression of the will of the executive that, through bureacratic, police, and military mechanisms, can quite readily be enforced.

At some point in new business discussions, LDC countries fall into two groups: some, like the PRC, want the essence of the business agreement left as an understanding among gentlemen; others, like Mexico, want everything in quintuplicate. But Mexico's paper mania should not be taken literally. In all but a few cases the real agreement among principals will not be reduced to writing; what will be written will be the minimum to satisfy government requirements.

It follows that lawyers do not perform the same functions in LDCs as they do in the U.S. and Western Europe. Lawyers are mainly management consultants, negotiators, and power-brokers. In these respects they function much like LDC bankers. Lawyers, therefore, see their role as defending their clients against possible sanctions by the government, while simultaneously feigning enthusiastic compliance with the law. Lawyers in LDCs are both management consultants and babysitters. They have to perform essential, mystical writing and editorial services at the conclusion of joint-venture discussions.

The police system in LDCs

As is widely known, police forces in LDCs are paid according to services rendered to the public they serve. The system is financed through two methods of taxation.

The one method is indirect, by means of tax revenues collected and redistributed in the form of wages and benefits. The other method is direct, by means of tax revenues collected at the premises or from persons, citizens and foreigners alike. In general, only the middle classes are so taxed; the upper classes are generally immune from taxation of whatever sort and the lower classes are too poor to pay.

LDC tax collectors-cum-policemen are generally conscientious about their work, and the ordinary citizen is ill advised to interfere with these routine processes of judicial/financial administration.

Corporate ethics in LDCs

To be born and raised in an LDC is one thing; to carry out business in an LDC as a foreigner is another. The difficulty is not that of not knowing anyone but of knowing the operating assumptions, the software, of government-industry relationships at the senior management level. The record of the performance of U.S. and European oil companies in Mexico during the 1920s and '30s stands as a monument to ignorance of the subject of government-industry relations in LDCs.

A crucial assumption lies in the area circumscribed by President Carter's Foreign Corrupt Practices Act. While the born-and-raised motorist of Mexico City will easily tip a patrolman for performing his duties in pulling the motorist over, the visiting tourist will generally find it difficult to pay a policeman in lieu of being taken, at once, to the police station for a moving traffic violation. The company operating in LDCs must pay the piper in many ways, but in some accounting categories managers can't be sure of who is going to end up with the money or (worse) who will have effective control of a local partner's share in the proposed joint venture. When in doubt, and before letters of intent are signed, a good excuse—corporate, regulatory, or even personal—which saves faces on both sides, is needed to exit gracefully.

Afterword

Much of this book has dealt with the phenomenal growth in Mexico of oil production, export capacity, and influence on the international oil market. The nationalistic, partly rhetorical, stance of Pemex and the Mexican government has been interpreted as the cross product of elements that are uniquely Mexican with elements that are common to LDCs in a general way and to LDCs who are Great Power neighbors in a particular way. Further, some effort has been made to look into the political and business psychology of Mexico's oil sector.

With its new development/recovery plan for 1983–1988 the de la Madrid government will play its oil card, changing the rules and structures of Pemex, its subsidiaries, contractors, and suppliers. The smokescreen dimension of the new qualitative program will succeed in making "some of the people," as Lincoln said, lose sight of the nuances and lessons of the López Portillo era. Each successive presidential administration can be expected to act in this way. Two threads running through this process are the relationship of Mexico to other oil exporters and the relationship, called, cryptically, the "special" one, between the U.S. and Mexico.

Subtle and dramatic changes

The 1976–1982 period in Mexico witnessed the full cycle of boom and bust, above all in oil industry. The financial crisis on the Mexican economy led to a dramatic contraction of trade for U.S. exporters to Mexico. Total U.S. exports to Mexico in 1982 dropped 32% from their 1981 level. U.S. exports to Mexico were $15.1 billion in 1980 and climbed to $17.4 billion in 1981. Then, in the financial crunch hitting Mexico, U.S. exports fell to $11.8 billion in 1982.[1]

A change in style in the post-oil boom period is the explicit and open role that the president of Mexico is taking in the oil industry. The tortuous and often hidden trail between Los Pinos and Pemex headquarters during the López Portillo tenure in office is now a four-lane freeway. This change can be seen in the way Mexico responded to the OPEC price reductions of March 1983. The government scheduled a session of the so-called Economic Cabinet on March 14 and promoted the affair as the meeting of economic statesmen, with the president of Mexico sitting as the de facto chairman of Pemex.

[1]*Mexican American Review,* March 1983, p. 13.

Attending the cabinet meeting were the Secretaries of Finance and Public Credit, Jesús Silva Herzog; of Budget and Planning, Carlos Salinas de Gortari; of Energy, Mines, and Parastate Industry, Francisco Labastida Ochoa; of the General Accounting Office, Francisco Rojas Gutiérrez; of Commerce and Industrial Development, Hector Hernández Cervantes; and of Labor and Social Welfare, Arsenio Farell Cubillas; as well as the presidents of the Bank of Mexico, Miguel Mancera Aguayo; Pemex, Mario Ramón Beteta; and the head of the Cabinet Secretariat, José Gamas Torruco. Reviewing the price and production decisions of OPEC, the government commented, with no hedging or rock-throwing, that OPEC's move "signifies a new commercial situation to which Mexico must adapt itself." How, in a word, would Mexico react? Responsibly—respecting the needs of both producers and consumers.

The government's Committee on Petroleum Exports, chaired by the head of Pemex and made up of the deputy secretaries of five ministries, presented alternatives for President de la Madrid's approval. The government press release emphasized the president's personal involvement in setting Pemex's new oil prices of $29/bbl for Isthmus and $23/bbl for Maya. The sleeper was that Mexico would not lower its export production level, which had been set since 1980 at 1.5 million b/d.

The government noted that the loss of revenues from oil exports would be made up, almost in their entirety, by savings in interest payments brought about by the trend toward lower rates. A savings of $1.5 billion was already in sight.

The press release was at pains to tie the president directly and intimately to Mexico's overall international petroleum policy: "President de la Madrid has cooperated in maintaining stability in the petroleum market and contributed to avoiding sudden changes in prices that would damage the international economy. . . . In summary, Mexico has acted in a serene, responsible, and coordinated fashion with the countries interested in achieving stability in the international petroleum market."

The mood of serenity surrounding the policy decisions of President de la Madrid and his Economic Cabinet is in sharp contrast to the hectic and turbulent environment permeating the oil boom years.

Recalling the mood of the oil boom

Mexico's oil sector during López Portillo's time was a shining export market for U.S., European, and Japanese suppliers of oil-field equipment and services. The oil sector's performance was behind the largest commercial loans in banking history. Non-Middle East oil contracts were made available to three dozen countries. In the process Mexico became the third largest trading partner of the U.S.—counting the trade in goods alone.

All of this created thousands of new relationships, personal, commercial, and economic. Manufacturers and bankers from all over the world sought to set up shop in Mexico on one basis or another. For bankers the formalities were easy: law prohibited them from commercial or retail banking, so their only open move to create a corporate presence in Mexico was to open a representative office in Mexico City. In 1979 alone over twenty Japanese banks opened their doors in Mexico.

Oil-field equipment manufacturers and services firms will tell a different story. In many instances they operated as pure exporters—and were paid in easy-to-count dollars. That, in the case of a manufacturer, his product was resold by his Mexican distributor for two or three times the export price—well, everyone adding value is entitled to his markup. The value added by the distributor was not worth a 200% markup? Or, put differently, if the market will support such an aggressive pricing strategy, why aren't we in the market ourselves?

This line of thought, motivated by common sense in most instances but by greed in others, led to a hornet's nest of issues and questions about establishing a corporate presence in Mexico's boom market. In the first place Mexico's foreign investment laws say that foreign capital in those industries not reserved to the state or Mexican capital can own up to 49% of the stock of a new company. Owning 49% implied that someone else owned 51%—but who? Foreign companies were told they needed joint-venture partners. "We don't need or want partners," was often their reaction. "Yes, you do by law," was the Mexican response. So, painfully, the matter of complying with Mexico's foreign investment laws began.

Banks, law firms, accounting firms, trade associations, nearly everyone got into finding the right partner and introduction business. One Mexican bank had a standard fee that it would quote for this service, but in general banks, Mexican and foreign, carried out such services as a professional courtesy. Since everything depended on personal and professional contacts, suddenly the fact of knowing someone in business or government had a potential cash value. Quite naturally, the introductions consultant regarded his collection of Mexican business cards as the equivalent of proprietary technology, and fees were charged for transferring this technology to clients or other consultants. Because the whole enterprise was being carried out with rules, ethical and commercial, that were still in the oven, on occasion ugly fights broke out among consultants who disagreed on the terms of trade. In one instance a U.S. consultant, serving as a go-between for a major U.S. oil company seeking a supply contract with Pemex, became so angry over what he perceived to be a half-hearted commitment by management to playing the introductions game that he threw up his hands and stormed out. Company management, meanwhile, could only wonder what all this righteous indignation was all about; the company was looking for crude oil contracts pure and simple. The consultant, for his part, had the idea that his business face in Mexico was in danger: what would happen if it were known that he was associated with a company whose managers refused to convert overnight to the doctrine of "corporate commitment to Mexico"?

Another U.S. consultant felt that he had lost valuable time and face at Pemex for having introduced a refinery representative who turned out to be a ne'er-do-well. Another U.S. consultant, who had recommended a Mexican associate for his high contacts in Pemex, was told by U.S. management that, on their trip to Mexico City in 1981 to meet Pemex brass, this particular consultant had stood them up repeatedly for scheduled business meetings. And their trip produced a "glad-to-meet-you" meeting with a third-level bureaucrat at Pemex. "Flakes," management concluded, ruefully.

The whole country, in short, turned into a who-you-know party, but it was one to which guests brought switchblades as well as party hats and balloons. Meanwhile, business seminars in the U.S. on "How to do business in Mexico," "Joint ventures in Mexico," and "How to live and work in Mexico" were sponsored regularly by the American Management Associations, the U.S. Department of Commerce, the World Trade Institute of New York, The U.S.-Mexico Chamber of Commerce (USMCOC), the Council of the Americas, and others. At $200–700 a head, a lot of money exchanged hands before a firm even crossed the border into Mexico. The joke of these meetings, where Mexican government speakers were invited, was in the small print on the seminar brochure: an asterisk often appeared by the name of a high-ranking official—at the bottom of the page it read "Invited." It became almost de rigueur for senior Mexican government invitees not to show up. Toward the end of López Portillo's time, seminar sponsors dared not go farther than to list the speaker as "a representative from Pemex." The cynicism on this score extended to Mexican government officials assigned to U.S. posts; on one occasion one of these had to arrange a luncheon speaker from Mexico City. "I knew he wouldn't come and almost took the risk of leaving his name off the brochure," laughed the official afterward.

Foreign companies complained that everyone even faintly related to Mexico had his hand out. Consultants, on the other hand, complained that companies always wanted marketing work done on a contingency basis: "There will be a good commission for you if you get a sale."

The heavy hitters in Mexican industry—the Alfas, the Protexas, the Permargos—were continually approached by foreign companies wanting to be their "partners." The president of one of these companies commented in 1980 that in the fifties and sixties he built up his oil-field business alone—there was no interest in Mexico from foreign companies. While his company had in recent years concluded a small number of joint ventures with U.S. companies, his general attitude was unsympathetic to the pleas of Johnnie-come-latelies.

Another dimension of the oil boom era was the discovery of markets and suppliers outside the U.S. Mexico's oil industry bought an immense amount of pipe from Japan, for example. Regarding joint ventures, Mexican companies by and large stayed with U.S. and European companies. The president of a Mexican pipe company commented "Why should I consider a joint venture with a Japanese company? We know how you Americans think, how and why you operate, what your language and religion are all about. As for Japan, I don't know their language, business ways or, really, anything that would give me confidence about the strength of my relationship." As a result of this lack of trust, which must have been evident from the Japanese side, many Japanese-Mexican joint ventures fell apart. (A notable exception to this discouraging pattern was a Japanese joint venture with a Mexican firm the principal of which was an American with a career's experience in Mexico's oil sector.)

Japanese businessmen, for their part, sometimes asked, What makes their Mexican counterparts tick? Were Mexicans, by character and outlook, essentially Americans who spoke Spanish? They also wondered, What is really going on in U.S.-Mexican relations? How do Mexicans really feel about Americans?

In any case, everyone in business who was new to Mexico's oil sector—Mexican, Japanese, U.S., or other—discovered that much of the industry, conceived of as a market for goods and services, is controlled from behind doors in Mexico City and Monterrey. The market was like the classical closed shop in which the union, the STPRM, in point of fact, controlled who got the jobs. That particular union didn't control all jobs, but the feel of the marketplace was that of an exclusive men's club whose members collectively formed a formidable oligopoly.

So much for mood in boom times. The mood during the Crash of '82 was one of anger, resentment, and disillusionment. "What became of your promises, our dreams?" asked middle-class Mexicans of the silent ghost of López Portillo. Other aspects of the oil sector, perhaps no less elusive, point to the future.

Trends in executive personnel selection at Pemex

The executive personnel system of the government's share of Mexico's oil industry is one such area of mystery. Senior managers of Pemex and related agencies come and go, on 2–6-year tours, very much as do officers in any military or diplomatic service. A tour at Pemex is likely to become a prerequisite for career officials aspiring to flag rank. But, as in the case of military careers, the invisible hand behind personnel assignments is never seen.

Tradition in Mexican government circles commands that at the end of a presidential sexennium, appointments by the new administration be kept absolutely secret. It is even said that a few decoy appointments are made, only to be undone on the day senior appointments are announced. Mexico City thrives on government gossip at most times of the year, but never so much as just before a new presidential administration is installed. The drama was so intense in 1982 that a proposal to shorten the presidential lame duck period (counted from the date of the announcement of the "tapped" nominee) for future elections was brought up.

Naturally, who is going to be named head of Pemex is one of the hotter topics of endless speculation. What can be known in advance about the identity of the new CEO of Pemex is that in all likelihood he will not be an oilman. In retrospect, López Portillo made a striking innovation in choosing Díaz Serrano, a businessman, petroleum engineer and trained historian, as the head of Pemex. Díaz Serrano's successors, renaissance-type administrators, represented a return to Mexican tradition in the area of executive appointments at Pemex.

Size of the industry

A second area of Mexico's oil industry laden with imponderables is the matter of the size of the oil sector. What is the total value added by all Mexican companies involved as suppliers and distributors of petroleum products, oil-field equipment and related services? How much in annual taxes do these companies pay? What is their total employment? What is their aggregate contribution to national income? Certainly it is more than the dogmatic 6%, but is it 12%, 18%, or 24%? No one seems to know or at least to be free to say.

Some oil subsidies are almost impossible to quantify. Millions of lower-class passengers ride 50 miles or more daily on Mexico City's subways for a fraction of a cent. Total revenues from fares probably fail to pay for vehicle and system maintenance, let alone electricity costs, including residual fuel. Pemex, too, is not charged for its energy consumption, thereby overstating its operating income by the value, at world prices, of the implicit subsidy.

The size of Mexico's oil industry is imponderable in another way. Over 95% of Mexico's gasoline service stations are privately owned but are operated by government concession holders. What are the politics and economics of the Pemex concession business in Mexico? Here, as in most areas of economic and management research related to the oil sector, Mexican academics and journalists are silent.

Financial and management reporting

Management control of the oil sector by means of financial accountability is certainly practiced in the government portion of Mexico's oil sector. During the López Portillo period, the statistical portions of the President's annual message to Congress were filled with tables showing management objectives by unit of measure, results, and percent variation. The IMP manual gives guidelines for financial reporting, and Pemex loan prospectuses give one or two clues about financial practices. But here too an invisible hand seems at work.

The behavioral results, as with the movement of the hands of a pocketwatch, are known, but a knowledge of the workings of the mainspring seems eternally elusive. That there is a system at work behind the face of the clock is certain, but it is doubtful that the outside observer, Mexican or otherwise, will be able to formulate a general theory of oil sector management in Mexico. The Mexican system wants it this way but insists that this veil should not be considered an obstacle to effective business relationships.

The backlash of success: vulnerability to world market fluctuations

The press release issued after the March 14 Economic Cabinet meeting concluded by saying that "Mexico is convinced that no single country can assume sole responsibility for controlling the market, yet neither can any nation evade its own responsibility in that regard. Mexico has assumed its responsibility." Because Mexico's policy had had a moderating effect on the policy of other exporters, it was important, the government observed, "not to underestimate the importance of Mexico in the international petroleum market."

This importance has also led to a higher degree of vulnerability to worldwide economic conditions. One quintessentially Mexican expression once described 99% of the business cycle in Mexico: "When Uncle Sam catches a cold, Mexico catches pneumonia." One intended result of the López Portillo program is that this expression is no longer valid in the way it used to be. One of the little-advertised blessings of Mexico's crude oil reserves was the promise of reducing Mexico's vulnerability to downward changes in the U.S. economy. Mexican oil policy succeeded in opening up new, non-U.S. markets, and oil served as the carrot that enticed otherwise reluctant countries and foreign companies into joint manufacturing ventures and technology-sharing agreements. The unintended result is that while this policy expanded Mexico's foreign markets and trade relations, it also internationalized Mexico's vulnerability to fluctuations in world market conditions.

It wasn't just a U.S. "cold" that hit Mexico in 1981–1982. Economic recessions in the industrialized world, cumulative impact of a decade of energy conservation measures, and Saudi Arabia's overproduction policy of 1979–1980 all combined to kill the growth in demand for crude oil in the industrialized countries. Mexico had increased production and export capacities with the implicit understanding underwriting the massive bank loans to Pemex that there would continue to be at least a U.S. market for Mexican crude oil.

Despite continued U.S. demand for Mexico's oil exports and a crucial financial boost from Uncle Sam's advance payments on oil deliveries to the S.P.R., Mexican vulnerability had become greater than the ability of U.S. assistance to pull her off the financial reef.

In Mexico, U.S. financial assistance was appreciated, but Mexico saw that help was far from all-out. In May 1983 the U.S. market was still closed to Mexican tuna, and Mexico was the only country worldwide whose tuna was on the embargo list. Commerce Secretary Hector Hernandez was forced to ask:

Given that Mexico is the number three trading partner of the U.S. and is its number one supplier of oil, is there no way for the U.S. government to suspend, while Mexico is undergoing this temporary critical period, the application of countervailing duties against Mexican exports alleged to be subsidized.[2]

A U.S. Department of Commerce official was asked at a San Francisco meeting in December 1979 of the U.S.-Mexico Chamber of Commerce, What is U.S. trade policy toward Mexico? He answered:

There is no special trade policy for Mexico. On the contrary, the U.S. is trying to internationalize and universalize the principles of trade relationships. For this reason the U.S. government hopes that Mexico 'will join the real world' by becoming a member of GATT.

The following March 18, 1980, Mexico announced its decision not to join GATT, thereby waving its right to have damage proven by export subsidies prior to a government's imposing countervailing duties. Mexico's commerce minister's request that the U.S. government not apply countervailing duties was, therefore, not as a matter of legal right, but of the nature of saying "please."

In his 1983 Cinco de Mayo speech in San Antonio, President Reagan, fishing for Hispanic votes, vowed to pay "special attention" to the economic consequences of Mexico's currency devaluations and debt crisis. The President said that his administration is "trying to do everything we can to work with Mexico itself in attacking the problem. . . . This is not just your problem, it's our problem and we'll meet it together."[3]

A long-standing rule in U.S.-Mexican relations is that only the tip of the iceberg can be made visible. The governments of the two countries behave quite formally toward each other, basing their relationships on sets of generalized principles. The very mention of a "special" relationship, therefore, is to hint at a topic that will never be officially explained or even recognized. The special quality can be seen only indirectly, in the advanced payments on S.P.R. deliveries, in the implicit U.S. government guarantee on loans from U.S. banks to Mexico, and in the official support given to the IMF in restructuring and refinancing Mexico's foreign debt. The special relationship means that for the U.S., Mexican political stability is very much a matter of national security, comparable in importance to issues like NATO and Japanese rearmament.

Mexico and the U.S. predictably disagree on matters of political style and content; they even disagree on what the issues are, to say nothing of possible solutions. And this is likely to be a permanent feature of the geography of U.S.-

---

[2]*Excelsior*, April 19, 1983.
[3]*San Francisco Chronicle* (May 6, 1983), p. 17.

Mexican relations. This cast-in-granite nonconvergence can be seen in the approaches of the two governments to the problem of political and military unrest in Central America. Both the U.S. and Mexico presumably have the same basic foreign policy goal: block Castroite imperialism before it spreads to Mexico. On how to accomplish this goal, the two countries differ widely, with Mexico anxious to limit both U.S. and Cuban involvement in the region by asserting diplomatic hegemony in the area.

Other countries are keen on Mexico and its long, porous border with the U.S. The USSR's embassy in Mexico has, or is close to having, a diplomatic staff larger than any other Russian embassy in the world. Mexico provides an excellent base for gathering industrial and military intelligence in the U.S. and Central America.

The importance to the U.S. firm of the special relationship between the U.S. and Mexico is that it has a direct impact on the volume of business transacted between the two nations. Would a major U.S. bank have loaned Pemex $100 million in early 1983 unless the U.S. Exim Bank were backing the deal? That's $100 million in business for U.S. suppliers of goods and services to Pemex and its Mexican oil-field contractors that would not have existed. How many more Mexican purchase contracts for U.S. goods would have been cancelled during 1982 without the U.S.'s advanced payments on S.P.R. deliveries? These examples illustrate a trend in international trade found all over the world: the supplier furnishes not only goods but arranges for the credit as well. In developing countries, including Mexico, cash-on-the-barrel days are receding.

How the special relationship affects business

During the Franco-Mexican oil and trade war of the summer of 1981, Mexico *insisted* that oil trade between Mexico and another country was evaluated in the context of the overall trade, investment, and technology-transfer relationships. Oil trade, in short, did not exist in a commercial vacuum—like it or not. American firms with their commercial sights on Mexico will therefore ignore the importance of the special relationship between the two countries at their risk. They may find, as did one manufacturer, that their Mexican distributor assumed that they understood these elementary facts of life.

When oil demand and prices dropped in 1981–1982, the result was economic pneumonia in Mexico: a massive loss of confidence in Mexico's private sector leading to a series of devaluations, exchange controls, cancellations of import permits, delays in payment by Pemex of 60–90–120 days, and fear and uncertainty on all sides.

Pemex's local and foreign suppliers reacted predictably. Management's marching orders were "close your line of credit, demand advance payments in cash or irrevocable letters of credit, and get outstanding invoices paid at once—before things get any worse."

In one case a San Francisco area electronics company that by no means considered itself in the oil business was an indirect vendor to Pemex. Through the company's Mexican distributors, Pemex had acquired and was quite satisfied with a number of this company's systems. Sales had been so good that the U.S. company and its Mexican distributors were convinced to enter into joint venture and technology transfer agreements to assemble the product in Mexico. These agreements were scheduled to close in June or July 1982, the Mexican company having agreed to pay $500,000 as a technology transfer fee. When the devaluation hit in August, the agreements had not been signed. The Mexican distributor and partner-to-be was suddenly faced with having to pay, in pesos, double the amount that the contract had implied to obtain the technology. When the exchange and import controls came in September, the U.S. company wondered if it was going to be possible even to export to Mexico, let alone assemble its products there.

Then came the telephone call, in late October. The conversation went something like this:

> *MEXICAN DISTRIBUTOR: We have an order from Pemex for an additional 10 systems. Get them to the border for us by this Saturday. Afterwards, our import permits aren't valid.*
> *U.S. SUPPLIER: The value of the shipment is $50,000. We've always worked on a letter of credit before. How are you going to pay for it?*
> *DISTRIBUTOR: Don't worry. We'll pay for it.*
> *SUPPLIER: How? With the exchange controls in effect?*
> *DISTRIBUTOR: I can't talk about it on the telephone. But believe me, we'll get the money to you within 3 weeks.*

The U.S. company wondered, What was this all about, not being able to talk about conditions of payment on the telephone? If the company's being paid depended on Pemex's paying, what were the chances of nonpayment or excessive delay in payment? The supplier's first reaction was yes (to the relief of the Mexican distributor), the shipment can be delivered to the border by Saturday. After the call was over, supplier management changed its mind. Their Mexican distributors, furious, had to tell Pemex that the sole source bid had to be withdrawn.

This incident provoked a deep and probably permanent rupture in the relations between the U.S. supplier and his Mexican distributor and partner-to-be. The Mexican distributor plans to look elsewhere for technology agreements, while the U.S. supplier concluded that its assembly program was a dead letter.

The point of the story is, first, that economic crises in Mexico characteristically start (and stop) in the U.S., which means that U.S.

suppliers and crude buyers have—like it or not—more than arms-length vendor and customer relations. This is the perpetual rub in U.S.-Mexican commerce. Mexicans expect treatment corresponding to their special relationship to the U.S., while Americans in business and government are likely to ask, *What* special relationship? when waters are stormy.

The incident between supplier and distributor also illustrates a point going beyond things U.S.-Mexican. In Mexico, as in LDCs in general, customer and supplier loyalty is highly valued. Commercial relations between companies and countries are taken personally by Mexican managers and government officials. An offense given to a Mexican company is felt personally by its owner and senior managers, who in turn are not likely to be very forgiving. In contrast, loyalty in the face of adversity is certain to be remembered for decades.

These two sides of Mexico's oil sector managers were discovered by a number of the U.S. major oil companies in 1979 and 1980 when, in the face of the Iranian crisis that had cut off oil supplies, they turned to Pemex to make up for supplies not being lifted in Iran. In several instances Pemex's response to the requested crude contract was something like this:

> PEMEX: *Thank you very much for coming all the way to Mexico City to see us. Recalling several previous incidents in which your company either cancelled or cut back on contracts similar to the ones you are now requesting, we find that at present all of our available export crude is under contract. Regretfully, therefore, we cannot accommodate you at the present time.*

U.S. oil companies could only ask themselves if Pemex's response was implying cause and effect. Oilmen sometimes heard the story told by a U.S. banker about how his firm had established close working relations with Pemex as far back as the 1940s when Pemex was still being blacklisted by the U.S. oil industry, in retaliation for the expropriation of oil properties in Mexico. The harvest from the seeds planted by the bank in the 1940s was being reaped in the bumper crop of the 1970s.

To a greater or lesser extent, the situation of U.S. oil companies that went to Pemex at the time of the Iranian crisis is the normal situation of all suppliers and buyers doing business in the Mexican oil sector. The oil sector is different from other Article 27 industries in Mexico due to the historical and ideological importance of oil industry politics.

From an outsider's perspective there is, therefore, what seems to be a strong, nonrational (not necessarily irrational) current underlying business relationships in Mexico's oil industry. At one moment a classic old-boy network. At another moment it seems as if any door—save the front door—is the way to do business with Pemex. Then, as if nothing out of the ordinary happened, a contract is signed, a special favor is granted, or a call is received at home on Sunday from a Pemex official who asks for advice about how Pemex should present its case to senior officials in the Mexican government. At such moments, the battle for management nerve and persistence seems worth fighting for.

Appendix A

Mexico macroeconomic indicators

Table A.1
Washing out currency overvaluation from peso exchange rate

	1973	*1974*	*1975*	*1976*	*1977*	*1978*	*1979*	*1980*	*1981*
Price levels									
Mexican prices									
1960 = 1.0	1.75	2.17	2.532	3.081	4.069	4.807	5.802	7.466	9.459
1981 = 1.0	0.19	0.23	0.27	0.33	0.43	0.51	0.61	0.79	1.00
U.S. prices									
1960 = 1.0	1.538	1.673	1.828	1.923	2.035	2.184	2.369	2.582	2.82
1981 = 1.0	0.55	0.59	0.65	0.68	0.72	0.77	0.84	0.92	1.00
Parity index									
1960 = 1.00	1.14	1.30	1.39	1.60	2.00	2.20	2.45	2.89	3.35
1981 = 1.00	0.34	0.39	0.41	0.48	0.60	0.66	0.73	0.86	1.00
Exchange rates									
1960 rate	12.50	12.50	12.50	—	—	—	—	—	—
Average	12.50	12.50	12.50	15.44	22.58	22.77	22.81	22.95	24.51
In 1960 pesos	7.14	5.76	4.94	5.01	5.55	4.74	3.93	3.07	2.59
Parity rate*	14.22	16.21	17.31	20.03	24.99	27.51	30.61	36.14	41.93

*The parity (of purchasing power) rate for a given year is obtained by taking the exchange rate of the previous year, multiplying it by the Mexican price level index (deflator) for the current year, and dividing the product by the U.S. price deflator for the current year. This method only adjusts for inflation differences. In the last year of a presidential cycle in Mexico, the demand for dollars in Mexico would exceed the level implied by mere inflation rate differences. The artificiality in using parity exchange rates retroactively comes about because the observed growth and inflation rates would have been different had the parity devaluation schedule been followed. In Pemex's case, using the parity rate gives an estimate of the degree to which Pemex imports were subsidized by other sectors of the economy and the degree to which the profitability of Pemex's operations measured in comparable units of purchasing was understated using the official exchange rates.
Source: Diemex-Wharton; parity rate adjustments by Baker & Associates

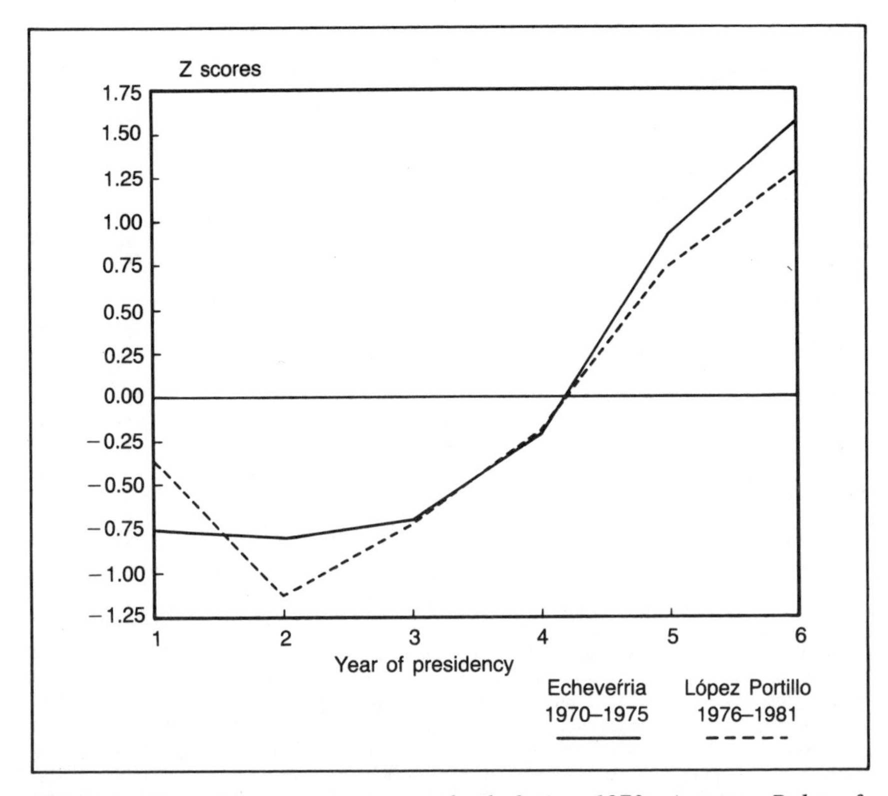

Fig. A.1 *How the peso was overvalued during 1970s (source: Baker & Associates)*

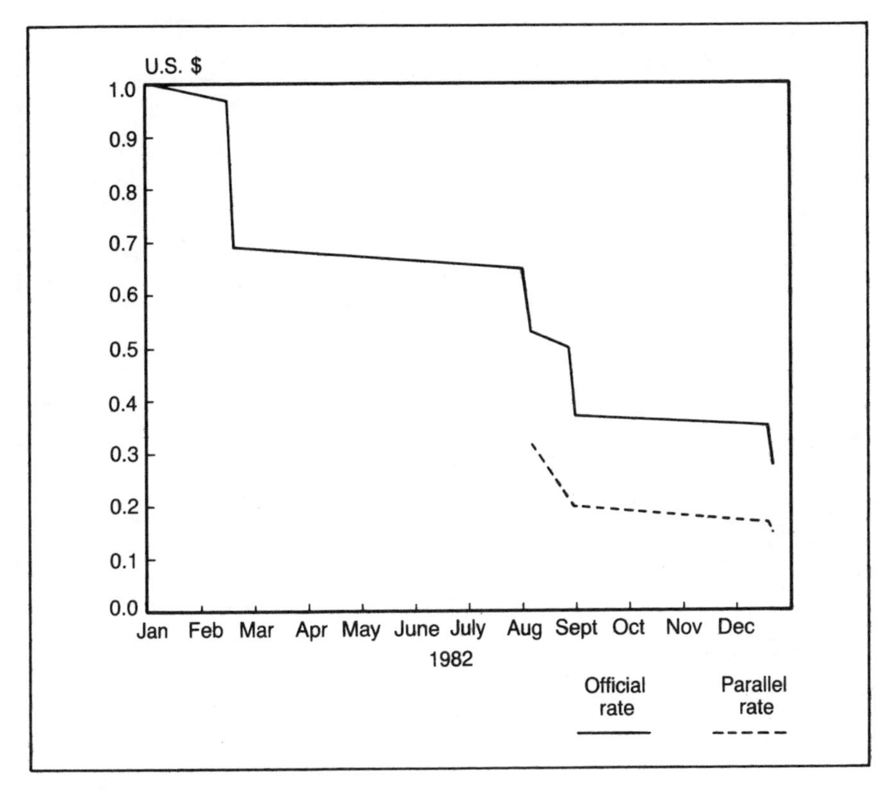

Fig. A.2 *How peso declined against U.S. dollar in 1982 (26.23 pesos upon remittance in January), (source: Baker & Associates)*

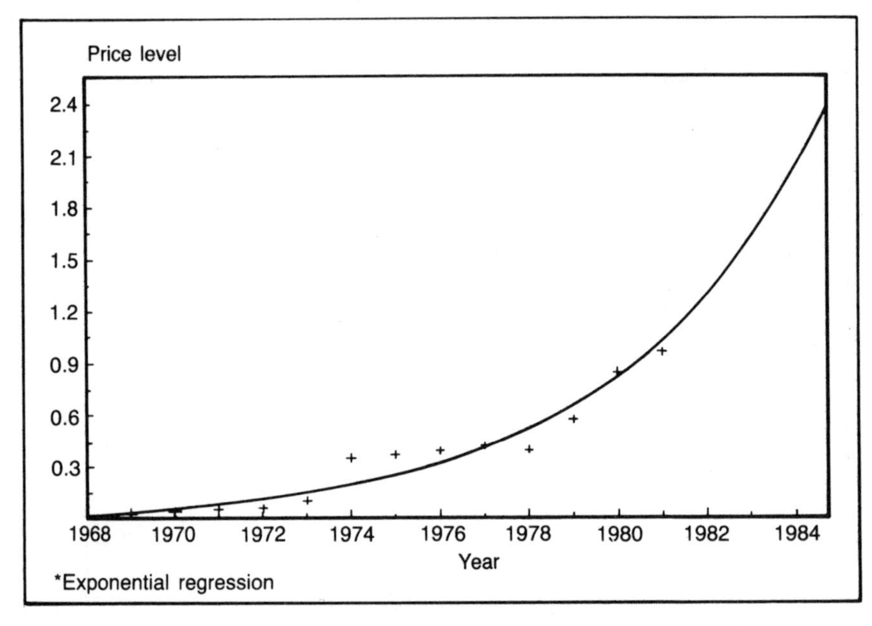

Fig. A.3 *Soaring oil industry inflation* (base year 1981 = 1.00), (source: Diemex-Wharton)*

Appendix B

Pemex financial management

Table B.1
Reconstructing Pemex annual operations

Spreadsheet line number*		Year ending December 31 (Million pesos)						Percent variation		
		1976	*1977*	*1978*	*1979*	*1980*	*1981*	*1977–81 Total*	*1977–81 Average*	*1981–80 Annual*
	Revenues									
14	Exports	6,835	23,724	41,898	92,876	239,136	349,284	1,372	96	46
15	Domestic	39,634	53,006	58,696	73,177	96,325	111,640	111	20	16
16	Net sales	46,469	76,730	100,595	166,053	335,461	460,924	501	57	37
17	Cost of goods sold	30,850	39,707	52,288	74,842	133,268	151,265	281	40	14
18	*Gross profit*	15,619	37,022	48,307	91,211	202,193	309,659	736	70	53
19										
20	Operating expenses									
21	Administration	NA	NA	NA	NA	NA	NA	—	—	—
22	Selling expense	4,657	7,896	9,995	15,763	17,582	29,937	279	40	70
23	Financial cost‡	645	2,298	4,199	12,182	4,327	980	−57	−19	−77
24	Total expenses	5,302	10,194	14,194	27,945	21,910	30,917	203	32	41
25	*Operating profit*	10,317	26,829	34,113	63,266	180,283	278,742	939	80	55
26	Financial and other									
27	expense and income									
28	Interest expense‡	841	773	556	1,503	10,900	32,194	4,065	154	195
29	Exchange losses	NA	NA	NA	−927	2,740	2,839	—	—	4
30	Other expense	249	6,801	3,826	372	1,783	3,149		−18	77
31	Other income†	−800	−936	−1,235	−1,272	−4,767	−4,757	408	50	−0.21
32	Net other	290	6,638	3,147	−323	10,656	33,425	404	50	214
33	*Pretax income*	10,027	20,190	30,966	63,590	169,627	245,317	1,115	87	45

Table B.1 Continued

*Spreadsheet line number**		*1976*	*1977*	*1978*	*1979*	*1980*	*1981*	*1977–81 Total*	*1977–81 Average*	*1981–80 Annual*
		Year ending December 31 (Million pesos)						*Percent variation*		
34										
35	Federal taxes	9,661	19,765	30,258	62,887	168,675	244,179	1,135	87	45
36	Net income	366	426	708	703	952	1,138	167	28	20
37	Employee profit-									
38	sharing	64	90	221	337	553	705	682	67	27
39	*Net profit*	303	336	487	366	398	433	29	7	9

*For Tables B.1–B.5
†Assumed net of expenses
‡Total financial cost aggregating interest, leasing and various taxes was:

	1976	1977	1978	1979	1980	1981
Total financial	1,486	3,071	4,755	13,685	15,227	33,174
Noncapitalized interest expense	841	773	556	1,503	10,900	32,194
Charged to operations	645	2,298	4,199	12,182	4,327	980

Source: Pemex, *Memoria de labores,* 1976-1981; Pemex financial statements

Table B.2
Pemex assets

Spreadsheet line number		Year ending December 31 (Million pesos)						Percent variation		
		1976	1977	1978	1979	1980	1981	1977–81 Total	1977–81 Average	1981–80 Annual
	Current assets									
62	Cash	5,712	2,751	4,841	6,344	6,564	14,820	439	52	126
63	Accounts									
64	receivable	7,636	14,617	12,351	13,714	36,357	83,960	474	55	131
65	Inventories									
66	Crude and products	NA	4,648	5,854	8,088	11,948	17,350	273	39	45
67	Supplies	NA	8,310	10,431	21,443	26,916	37,899	356	46	41
68	Inventory total	10,152	12,958	16,285	29,532	38,864	55,249	326	44	42
69	*Total current*									
70	*assets*	23,500	30,326	33,477	49,589	81,786	154,029	408	50	88
71	Investments									
72	Trust funds	NA	7,013	5,806	0	—	—	—	—	—
73	Nonconsolidated									
74	subsidiaries	NA	184	184	184	199	273	48	10	37
75	Other	NA	120	107	828	1,228	2,749	2,189	119	124
76	*Total*	9,137	7,317	6,098	1,012	1,427	3,022	−59	−20	112
77	Plants and equipment									
78	Land	NA	545	662	914	1,215	1,359	150	26	12
79	Buildings	NA	3,363	4,093	5,547	6,801	11,788	251	37	73
80	Wells	NA	29,440	34,379	41,152	58,745	85,284	190	30	45
81	Plants, equipment	NA	50,728	64,984	84,447	166,005	238,286	370	47	44
82	Pipelines	NA	16,164	20,769	41,209	60,227	95,676	492	56	59
83	Maritime fleet	NA	5,574	5,800	9,907	13,214	13,705	146	25	4

Spreadsheet line number		*Year ending December 31 (Million pesos)*						*Percent variation*		
		1976	*1977*	*1978*	*1979*	*1980*	*1981*	*1977–81 Total*	*1977–81 Average*	*1981–80 Annual*
84	Other	NA	416	415	414	414	413	−1	0	0
85	*Total*	45,431	106,230	131,101	183,589	306,620	446,511	320	43	46
86	Less:									
87	Depreciation	NA	49,041	59,995	71,986	86,883	113,803	132	23	31
88	Net	NA	57,189	71,106	111,604	219,737	332,708	482	55	51
89	Installations and									
90	construction									
91	in progress	32,022	52,877	100,045	124,966	154,230	248,444	370	47	61
92	Revaluation of									
93	property, wells,									
94	plants, and equipment	NA	132,573	136,116	136,116	280,171	442,925	234	35	58
	Net assets	110,090	280,281	346,842	423,288	737,351	1,181,128	321	43	60
95	Deferred charges	108	195	197	386	1,765	6,895	3,441	144	291
96	*Total assets*	110,198	280,476	347,038	423,674	739,116	1,188,023	324	43	61

Source: Pemex; *Excelsior*, Aug. 22, 1982

Table B.3
Pemex liabilities & capital

Spreadsheet line number		Year ending December 31 (Million pesos)						Percent variation		
		1976	1977	1978	1979	1980	1981	1977–81 Total	1977–81 Average	1981–80 Annual
110	*Current liabilities*									
111	Notes and bonds	NA	14,585	25,266	33,486	42,348	185,505	1,172	89	338
112	Suppliers and									
113	contractors	NA	13,960	25,674	28,974	47,575	37,704	170	28	−21
114	Accounts payable	NA	2,080	2,856	4,713	8,659	35,851	1,624	104	314
115	Accrued liabilities	NA	1,931	3,429	22,375	31,362	25,533	1,222	91	−19
116	Federal tax payable	NA	3,897	6,724	22,375	31,603	40,358	936	79	28
117	*Total current*	30,964	36,453	63,948	97,198	161,547	324,951	791	73	101
118	Long-term liabilities*									
119	Notes and bonds	NA	61,376	80,054	123,669	177,661	292,069	376	48	64
120	Suppliers and									
121	contractors	NA	1,422	12,686	3,978	6,083	6,972	390	49	15
122	Advances from									
123	customers	—	—	—	—	9,306	0	—	—	—
124	*Total long term*	37,894	62,797	92,739	127,647	193,050	299,041	376	48	55
125	Provision									
126	Retirement plans	NA	1,737	1,866	2,341	2,582	2,910	68	14	13
127	Sundry liabilities	NA	1,316	1,792	5,144	9,620	15,038	1,043	84	56
128	Total	2,483	3,052	3,657	7,485	12,202	17,948	488	56	47
129	*Total liabilities*	71,341	102,302	160,344	232,330	366,799	641,940	527	58	75
130	*Capital*									
131	Contributions of									
132	federal government	NA	6,318	6,318	6,318	6,318	6,318	—	—	—

Spreadsheet line number		Year ending December 31 (Million pesos)						Percent variation		
		1976	*1977*	*1978*	*1979*	*1980*	*1981*	*1977–81 Total*	*1977–81 Average*	*1981–80 Annual*
134	Specific reserves									
135	Equipment and instal-									
136	lation replacement	NA	2,216	2,216	2,216	2,216	2,216	—	—	—
137	Self-insurance	NA	332	311	311	311	311	—	—	—
138	Exploration and									
139	depletion	NA	34,226	38,837	43,156	78,147	87,260	155	26	12
140	*Total reserves*	30,400	36,774	41,363	45,683	80,674	89,787	144	25	11
141	Revaluation of									
142	property, etc.	NA	132,573	136,116	136,116	281,701	445,921	236	35	58
143	Prior years'									
144	surplus	1,900	2,174	2,509	2,896	3,226	3,624	67	14	12
145	Net profit	303	336	487	366	398	433	29	7	9
146	*Total capital*	38,837	178,174	186,694	191,344	372,317	546,083	206	32	47
147	*Total liabilities*	110,198	280,476	347,038	423,674	739,116	1,188,023	324	43	61
148	*and capital*									
149										
150	*Net worth*									
151	(Lines 96–128)	107,615	277,424	343,381	416,189	726,914	1,170,075			

*Pemex's annual financing program is approved by the Secretariat of Finance and Public Credit, but the Mexican government "is not, however, legally responsible for the obligations incurred by Pemex."
Source: Pemex syndicated acceptance facility; *Excelsior*, Aug. 22, 1982.

Table B.4
Low Pemex net jacks debt needs for working capital

Spreadsheet line number		Year ending December 31 (Million pesos)						Percent variation		
		1976	1977	1978	1979	1980	1981	1977–81 Total	1977–81 Average	1981–80 Annual
165	*Source*									
166	*Net income*	303	336	487	366	398	433	29	7	9
167	Reserve credits									
168	(Charged to cost of sales)									
169	Depreciation	NA	8,787	11,042	12,089	14,928	26,979	207	32	81
170	Exploration and depletion		6,409	4,611	4,320	34,991	9,113	42	9	−74
171	Self-insurance	NA	0	−22	0	0	0			
172	Retirement	NA	119	129	475	241	328	176	29	36
173	Sundry liabilities	NA	450	476	3,352	4,477	5,418	1,103	86	21
174	Depreciation/	NA	0	0	0	1,530	1,466	—	—	−4
175	property revaluation									
176	*Total operation*	NA	16,101	16,624	20,566	56,563	43,737	172	28	−23
177	*funds*									
178	Long-term debt	NA	27,636	47,813	62,833	81,460	154,229	458	54	89
179	Sale of noncurrent	NA	7	0	2,768	5,320	168	2,266	121	−97
180	assets									
181	Total	NA	43,744	64,437	86,166	143,343	198,134	353	46	38
182	Applications									
183	Noncurrent assets	NA								
184	(net)	NA								

Spreadsheet line number		*Year ending December 31 (Million pesos)*						*Percent variation*		
		1976	*1977*	*1978*	*1979*	*1980*	*1981*	*1977–81 Total*	*1977–81 Average*	*1981–80 Annual*
185	Trust funds	NA	−1,823	−1,206	−5,806	0	0	—	—	—
186	Property	NA	7,086	8,497	7,536	25,250	33,639	375	48	33
187	Installations in	NA	34,320	63,631	72,740	132,395	200,693	485	56	52
188	progress									
189	Deferred charges	NA	87	2	189	1,379	5,130	5,797	177	272
190	Investments	NA	3	−13	721	415	1,595	51,352	376	285
191	Devaluation adjustment	NA								
192	(Credited to long-	NA	−6,743	−5,362	−3,097	−3,348	0	−100	−100	−100
193	term debt)									
194	Long-term debt due									
195	within 1 year	NA	9,476	23,233	31,021	16,057	48,238	409	50	200
196	*Total*	NA	42,406	88,781	103,304	175,495	289,295	582	62	65
197	*In(de)crease in*									
198	*working capital*	NA	1,337	−24,344	−17,138	−32,152	−91,161	—	—	—

Source: Pemex

Table B.5
Pemex's shaky financial ratios

*Spreadsheet line number**		*Year ending December 31 (Percent or times)*						*Percent variation*		
		1976	*1977*	*1978*	*1979*	*1980*	*1981*	*1977–81 Total*	*1977–81 Average*	*1981–80 Annual*
	Solvency									
211	Current ratio	0.76	0.83	0.52	0.51	0.51	0.47	−43	−13	−6
212	(70/177)									
213	Quick ratio	0.43	0.48	0.27	0.21	0.27	0.30	−36	−11	14
214	(62 + 64)/117									
215	Quick ratio (investment)	NA	NA	NA	0.29	0.34	0.36	—	—	5
216	(62 + 64 + 66)/117									
217	Profit margins									
218	Gross (18/16)	33.61	48.25	48.02	54.93	60.27	67.18	39	9	11
219	Operating (25/16)	22.20	34.96	33.91	38.10	53.74	60.47	73	15	13
220	Pretax (33/16)	21.58	26.31	30.78	38.29	50.57	53.22			
221	Net (39/16)	0.65	0.44	0.48	0.22	0.12	0.09	−79	−32	−21
222	Current liabilities to net									
223	worth (117/151)	28.77	13.14	18.62	23.35	22.22	27.77	111	21	25
224	*Return on assets:*									
225	To Mexican govern-									
226	ment (33/96)	9.11	7.20	8.92	15.01	22.95	20.65	187	30	−10
227	To Pemex (39/96)	0.27	0.12	0.14	0.09	0.05	0.04	−70	−26	−32
228	*Return on net*									
229	*worth (equity):*									
230	To Mexican govern-									
231	ment (33/151)	9.32	7.28	9.02	15.28	23.34	20.97	188	30	−10
232	To Pemex (39/151)	0.28	0.12	0.14	0.09	0.05	0.04	−69	−26	−32

Table B.5 Continued

*Spreadsheet line number**		*Year ending December 31 (Percent or times)*						*Percent variation*		
		1976	*1977*	*1978*	*1979*	*1980*	*1981*	*1977–81 Total*	*1977–81 Average*	*1981–80 Annual*
233	*Efficiency indicators*									
234	Collection period	60	70	45	30	40	66	−4	−1	68
235	(days) (63/16 × 365)									
236	Net sales to									
237	inventory (16/65)	NA	17	17	21	28	27	61	13	−5
238										
239	Unit profits ($/bbl)									
240	Output (million									
241	bbl oil equivalent)	444	508	627	749	971	1,150	126	23	18
242	Average exchange									
243	rate (peso/$)	15.44	22.58	22.77	22.81	22.95	24.51	9	2	7
244	Net sales/bbl	6.78	6.69	7.05	9.72	15.05	16.35	144	25	9
245	Gross profit/bbl	2.28	3.23	3.38	5.34	9.07	10.99	240	36	21
246	Operating profit/									
247	bbl	1.50	2.34	2.39	3.70	8.09	9.89	323	43	22
248	Tax/bbl	1.41	1.72	2.12	3.68	7.57	8.66	403	50	14
249	*Net profit/bbl*	0.04	0.03	0.03	0.02	0.02	0.02	−47	−15	−14

*Spreadsheet line numbers appear with ratios

Source: Tables B.1–B.5; Dunn & Bradstreet, *Selected Key Business Ratios in 125 Lines of Business* (1981); Table 7.1

Table B.6
What Pemex imports for current operations

	(Million $)							
	1975	*1976*	*1977*	*1978*	*1979*	*1980*	*1981*	*1982**
Petroleum								
Products	226	126	52	144	209	243	159	120
Petrochemicals	57	104	157	164	332	523	524	318
Subtotal	283	230	208	308	540	766	683	438
Other								
Goods and services	434	547	486	1,009	1,070	NA	NA	NA
Technical assistance	NA	NA	NA	NA	NA	NA	NA	NA
Equipment leased	—	—	—	—	255	100	156	NA
Less: merchandise imports for Pemex contractors and other resale	NA	212	172	282	461	797	604	NA
Net Pemex other	NA	335	314	727	609	NA	NA	NA
Total Pemex imports for current operations†	NA	565	522.5	1,034.2	1,149.1	NA	NA	NA

*August
†Exclusive of the value of imported financial services.
Source: J. Corredor, "El petroleo en Mexico," Table 38; Pemex, *Memoria de labores,* 1976–81.

Table B.7
How overvalued peso affected Pemex margin

	1976	*1977*	*1978*	*1979*	*1980*	*1981*
Official average exchange rate*	15.44	22.58	22.77	22.81	22.91	24.51
Parity rate	20.03	24.99	27.51	30.61	36.15	41.93
Imports†	565	523	1,034	1,149	NA	NA
			(Million pesos)			
Official rate						
Exports	6,835	23,724	41,898	92,876	239,136	349,284
Domestic	39,634	53,006	58,696	73,177	96,325	111,640
Net sales	46,469	76,730	100,595	166,053	335,461	460,924
Cost of goods						
Imported goods and services‡	8,724	11,798	23,549	26,211	29,300	23,700
Other costs	22,126	27,909	28,739	48,631	103,968	127,565
Total	30,850	39,707	52,288	74,842	133,268	151,265
Gross profit	15,619	37,022	48,307	91,211	202,193	309,659
Gross margin (%)	34	48	48	55	60	67
			(Million parity-adjusted pesos)			
Parity rate						
Exports§	11,317	26,256	50,620	124,636	376,678	597,531
Domestic	39,634	53,006	58,696	73,177	96,325	111,640
Net sales	50,951	79,262	109,316	197,813	473,003	709,171
Cost of goods						
Imported goods and services¶	11,317	13,057	28,451	35,174	46,152	40,544
Other costs	22,126	27,909	28,739	48,631	103,968	127,565
Total	33,443	40,967	57,190	83,805	150,120	168,109
Gross profit	17,508	38,295	52,126	114,008	322,883	541,061
Gross margin (%)	34	48	48	58	68	76

*The devaluation of September 1976 caused the peso to go to 19.95:1$ from 12.5:1$; taking the annual average rate gives an estimate of Pemex's actual dollar revenues and import costs for 1976.
†Million $ (Table B.6).
‡1980 and 1981 are estimated.
§Correction for overvalued peso generates higher revenues.
¶Correction generates higher costs; values are net of imports for resale.
Source: Table B.6 and Pemex; adjustments for peso overvaluation by Baker & Associates; average exchange rates, Diemex-Wharton.

Table B.8
Pemex reports on its cash flow
Year ending December 31

	(Million pesos)					
	1976	*1977*	*1978*	*1979*	*1980*	*1981*
Inflow						
Income	50,774	78,336	113,310	184,372	361,600	471,773
Borrowings	21,254	27,635	51,306	74,654	134,000	397,963
Total	72,028	105,971	164,616	259,026	495,600	869,736
Outflow						
Current expense						
Operations	23,723	33,587	43,450	66,104	100,600	134,217
Interest expense	841	773	556	1,503	10,900	32,194
Other	4,261	3,483	5,862	5,780	9,300	30,901
Subtotal	28,825	37,843	49,868	73,387	120,800	197,312
Federal taxes	7,615	18,898	27,213	47,014	162,400	238,193
Subtotal	36,470	56,741	77,081	120,401	283,200	435,505
Capital Account						
Investments	21,335	34,916	62,703	83,472	121,800	230,773
Debt liquidation	12,237	13,821	18,380	40,805	70,500	165,657
Interest expense	2,578	3,678	6,862	15,006	19,800	29,546
Subtotal	36,149	52,415	87,945	139,283	212,100	425,976
Total outflow	72,589	109,156	165,026	259,684	495,300	861,481
Net flow	−561	−3,185	−410	−658	300	8,255

	(Percent)					
	1976	*1977*	*1978*	*1979*	*1980*	*1981*
Inflow						
Income	70.49	73.92	68.83	71.18	72.96	54.24
Borrowings	29.51	26.08	31.17	28.82	27.04	45.76
Outflow						
Current expense						
Operations	65.05	59.19	56.37	54.90	35.52	30.82
Interest expense	2.31	1.36	0.72	1.25	3.85	7.39
Other	11.68	6.14	7.60	4.80	3.28	7.10
Subtotal	79.04	66.69	64.70	60.95	42.66	45.31
Federal taxes	20.88	33.31	35.30	39.05	57.34	54.69
Capital Account						
Investments	59.02	66.61	71.30	59.93	57.43	54.18
Debt liquidation	33.85	26.37	20.90	29.30	33.24	38.89
Interest expense	7.13	7.02	7.80	10.77	9.34	6.94

Source: Pemex, *Memoria de labores,* 1976–1981.

Table B.9
A look at Pemex cost of sales
Year ending December 31

	(Million pesos)					
	1976	*1977*	*1978*	*1979*	*1980*	*1981*
Operating costs						
Oil fields						
Actual cost	11,357	13,514	15,907	23,906	NA	NA
Reserve charges	7,635	11,407	15,960	22,610	64,052	75,575
Subtotal	18,992	24,921	31,867	46,516	NA	NA
Refineries	6,765	8,768	9,925	13,373	NA	NA
Petrochemical plants	3,288	3,780	4,836	6,496	NA	NA
Provisions	1,219	1,579	—	—	—	—
Purchase of crude and refined products	3,679	4,003	8,178	1,291	NA	NA
Purchase of goods for resale	97	79	377	932	NA	NA
Subtotal	34,040	43,129	55,182	68,608	NA	NA
Less:						
Inventory variation	764	443	1,050	1,947	NA	NA
Products consumed in operations	1,207	1,400	1,845	2,619	NA	NA
Subtotal	1,971	1,843	2,894	4,566	NA	NA
Total cost of goods sold	32,069	41,286	52,288	64,042	133,268	151,265

Source: *La industria petrolera en México,* 1980, Table V.2

Table B.10
Who owes Pemex money

	(Million pesos)				
	1977	*1978*	*1979*	*1980*	*1981*
Local trade	4,367	5,272	5,106	12,387	14,211
Foreign trade	6,139	1,758	1,706	8,272	36,008
Government accounts	2,587	615	1,358	1,269	819
Tax credit	0	0	0	7,939	19,302
Employees and others	1,123	2,730	4,970	6,562	13,546
Advances to suppliers	542	1,947	673	234	887
Mexican government	161	276	161	270	496
Total	14,919	12,598	13,974	36,932	85,269
Less allowance for bad debts	302	247	260	575	1,309
Net	14,617	12,351	13,714	36,357	83,960

	Accounts receivable (Percent)				
	1977	*1978*	*1979*	*1980*	*1981*
Local trade	29.27	41.85	36.54	33.54	16.67
Foreign trade	41.15	13.96	12.21	22.40	42.23
Government accounts	17.34	4.88	9.72	3.44	0.96
Tax credit	0.00	0.00	0.00	21.49	22.64
Employees and others	7.53	21.67	35.56	17.77	15.89
Advances to suppliers	3.63	15.45	4.82	0.63	1.04
Mexican government	1.08	2.19	1.15	0.73	0.58
Less allowance for bad debts	2.03	1.96	1.86	1.56	1.54
Net	97.97	98.04	98.14	98.44	98.46

Source: Pemex

Table B.11
Pemex's 1982 pretax income jumps 175%

*Spread sheet line number**		*(Millions of current pesos)* 1982	1981	% Var.
	Revenues			
14	Exports	916,409	349,284	162
15	Domestic	162,573	111,640	46
16	Net sales	1,078,982	460,924	134
17	Cost of goods sold	277,901	151,265	84
18	*Gross profit*	801,081	309,659	159
19				
20	Operating expenses			
21	Administration	NA	NA	
22	Selling expense	58,248	29,937	95
23	Financial cost	NA	980	
24	*Total expenses*	58,248	30,917	88
25	Operating profit	742,833	278,742	166
26	Financial and other			
27	expense and income:			
28	Interest expense	71,603	32,194	122
29	Exchange losses	NA	2,839	−100
30	Other expense	3,974	3,149	26
31	Other income	−6,174	−4,758	30
32	Net other	69,403	33,425	108
33	*Pretax income*	673,430	245,318	175
34				
35	Federal taxes	671,930	244,179	175
36	Net income	1,500	1,138	32
37	Employee profit			
38	sharing	938	705	33
39	*Net profit*	562	433	30

*As shown in Table B.1
Source: Pemex, *Estados financieros al 31 de diciembre de 1982;* Table B.1

Table B.12
Pemex's income taxed at 99%

Spread sheet line number		*(Billions of current pesos)* *1982*	*1981*	*% Var.*
13	Net sales	1,079	461	134
17	Cost of goods sold	278	150	85
18	Gross profit	1,357	611	122
	% net sales	80	75	5
20	Operating expenses			
21	Administration	NA	NA	—
22	Selling expenses	58	30	93
34	*Total*	58	30	93
25	Operating profit	743	281	164
	% net sales	69	61	13
23	Financial costs	72	36	100
36	Other income/			
27	expense (net)	−2	2	−200
33	Pretax income	673	245	175
	% of net sales	62	53	
35	Federal taxes	672	244	175
	Tax rate (%)	99.85	99.59	
39	*Net profit*	1	1	—

*Table B.1
Source: Pemex, *Memoria de labores, 1982;* Table B.1

Table B.13
Pemex's foreign debt assumed in 1982

(Million pesos) *Lender*	*Amount*
Long term	
Citibank, N.A. (syndicate)	90,000
Banque de la Societe Financiere (agent)	2,820
Bank of America	2,560
Swiss Bank	1,074
Banque de Paris et des Pays Bas	1,410
Credit Suisse	1,150
Daiwa Securities	800
Commerzbank, A.G.	2,430
Swiss Bank Corp.	2,400
Citibank, N.A.	6,750
Subtotal	361,200
Short term	174,691
Total	535,891

Source: Pemex, *Memoria de labores, 1982*

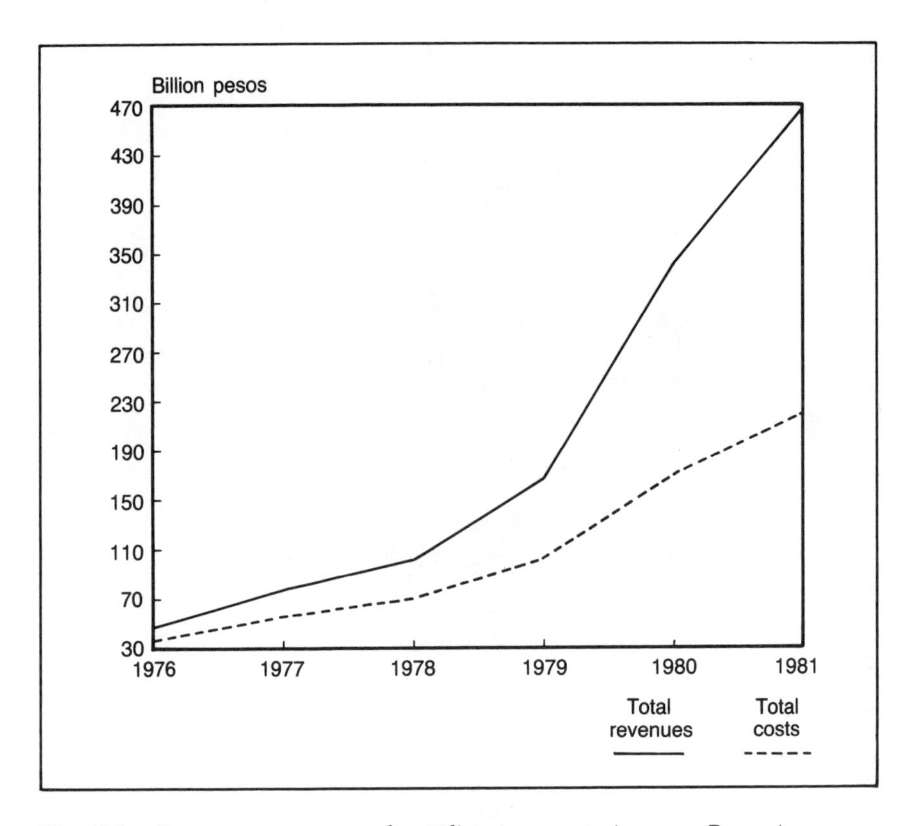

Fig. B.1 *Pemex revenue growth outdistances costs (source: Pemex)*

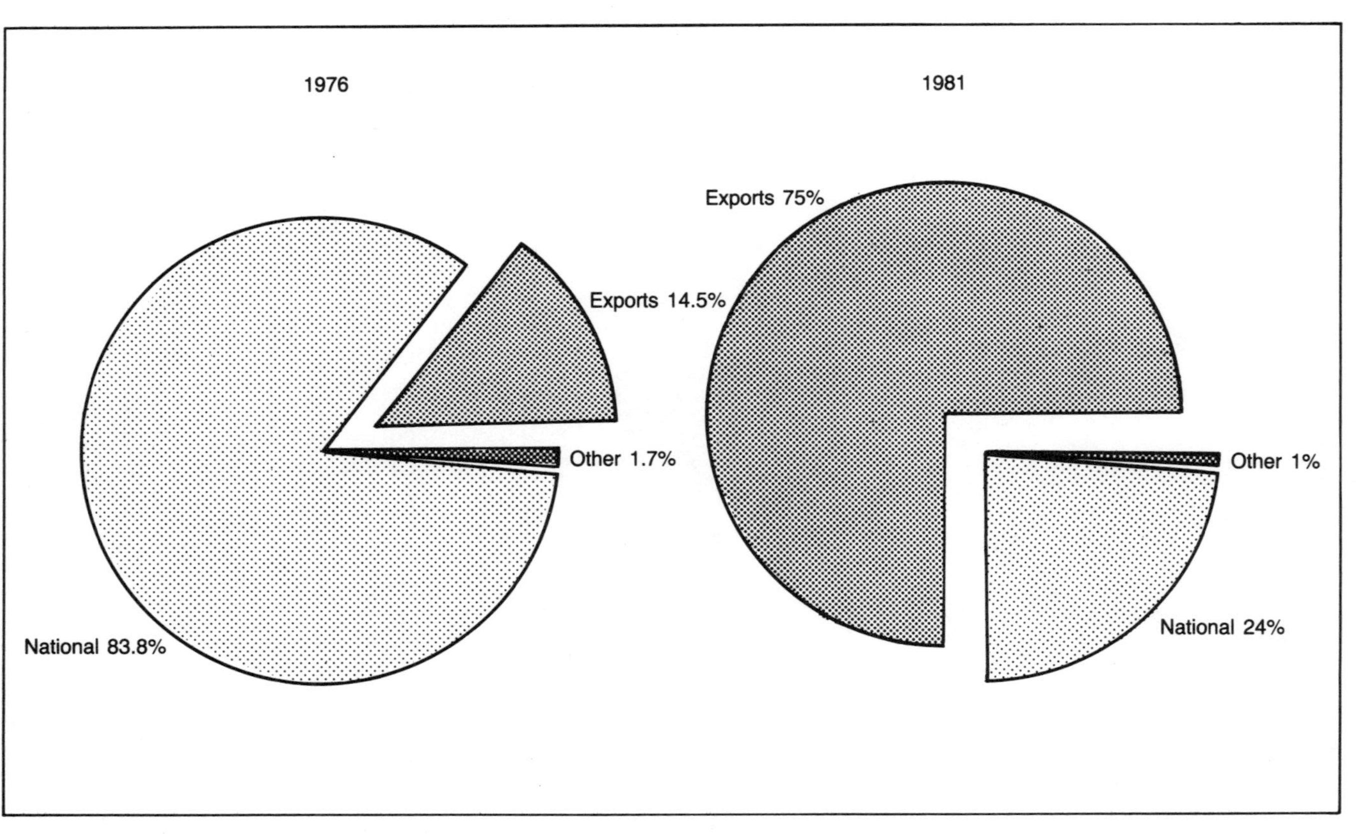

Fig. B.2 *Where Pemex revenues originate (source: Pemex)*

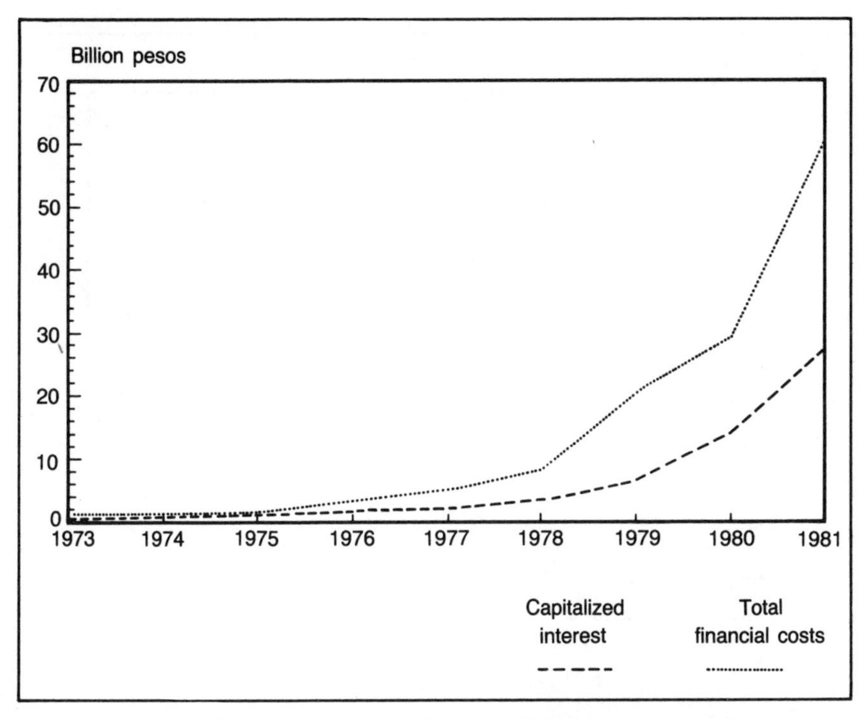

Fig. B.3 *Pemex financial costs rise (source: Baker & Associates)*

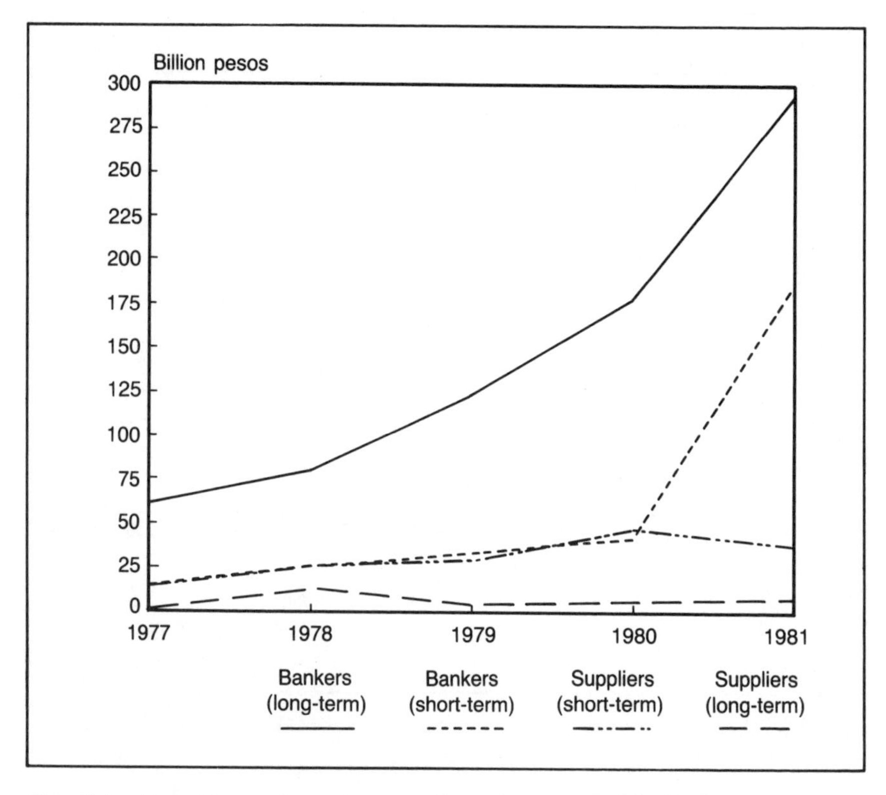

Fig. B.4 *How Pemex leaned on creditors (source: Table B.3)*

Appendix C

Investments by Pemex

Table C.1
Pemex investment program

	(Million $)							
	1974	*1975*	*1976*	*1977*	*1978*	*1979*	*1980*	*1981*
Exploration	46	72	112	81	62	145	150	482
Drilling								
Exploratory	110	123	130	142	218	286	688	849
Development	149	184	191	176	289	379	683	845
Total	259	307	322	318	507	665	1,370	1,694
Production	118	145	256	280	831	1,223	1,472	1,904
Refining	97	259	270	356	444	539	466	814
Petrochemicals	96	146	258	271	478	771	723	1,083
Transportation and distribution	95	164	222	176	291	168	373	845
Other	11	3	3	36	113	118	76	726
Total	721	1,095	1,444	1,517	2,726	3,629	4,629	7,547

	(Million pesos)							
	1974	*1975*	*1976*	*1977*	*1978*	*1979*	*1980*	*1981*
Exploration	576	900	2,248	1,858	1,435	3,324	3,440	11,599
Drilling								
Exploratory	1,376	1,533	2,604	3,268	5,015	6,582	15,813	20,436
Development	1,857	2,298	3,828	4,039	6,647	8,724	15,698	20,354
Total	3,233	3,831	6,432	7,307	11,662	15,306	31,501	40,790
Production	1,476	1,810	5,126	6,435	19,121	28,128	33,860	45,836
Refining	1,207	3,240	5,401	8,193	10,203	12,399	10,721	19,601
Petrochemicals	1,201	1,825	5,162	6,233	10,994	17,733	16,619	26,079
Transportation and distribution	1,185	2,049	4,441	4,041	6,682	3,857	8,568	20,564
Other	135	37	64	826	2,606	2,725	1,748	17,458
Total	9,013	13,693	28,874	34,893	62,703	83,472	106,457	181,737

Source: Pemex

Appendix D
Mexico's hydrocarbon reserves

Table D.1
Hydrocarbon reserves and production

(Million bbl oil equivalent)

Year	*Total reserves (year end)*	*Reserves index 1938 = 100*	*Total production*	*Production index 1938 = 100*	*Cumulative production*	*Reserves: production years*
1901–1937					1,874	
1938	1,276	100	44	100	1,918	29
1971	5,428	425	298	677	6,492	18
1972	5,388	422	309	702	6,801	17
1973	5,432	426	318	723	7,119	17
1974	5,773	452	379	861	7,498	15
1975	6,338	497	439	998	7,937	14
1976	11,160	875	469	1,066	8,406	24
1977	16,002	1,254	533	1,211	8,939	30
1978	40,194	3,150	658	1,495	9,597	61
1979	45,803	3,590	785	1,784	10,382	58
1980	60,126	4,712	1,015	2,307	11,397	59
1981	72,008	5,643	1,199	2,725	12,596	60

*Includes crude, condensate, natural gas liquids and natural gas. Pemex converts gas to oil equivalent at 5,000 cu ft/bbl.
Source: Pemex

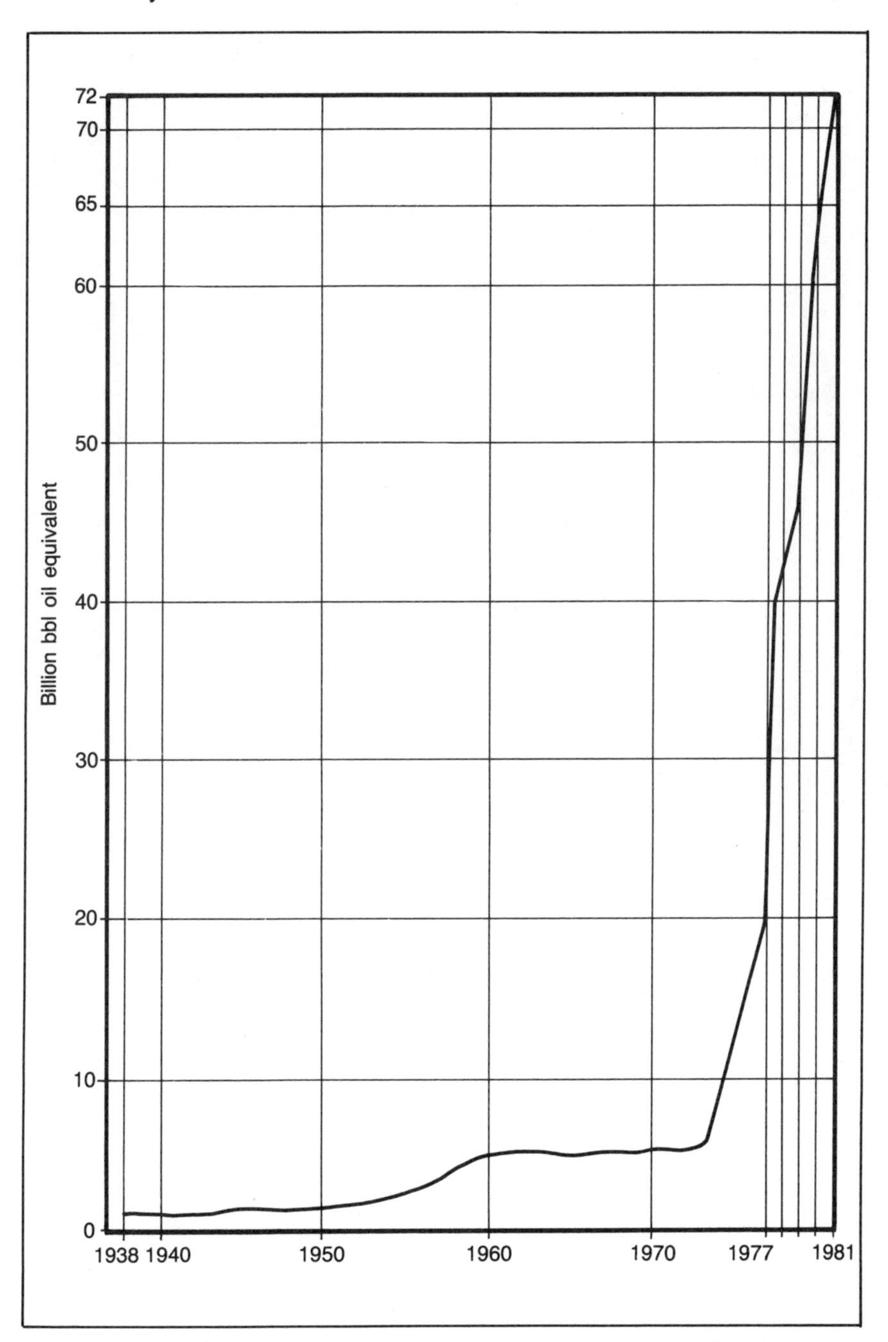

Fig. D.1 *How hydrocarbon reserves increased (source: Pemex)*

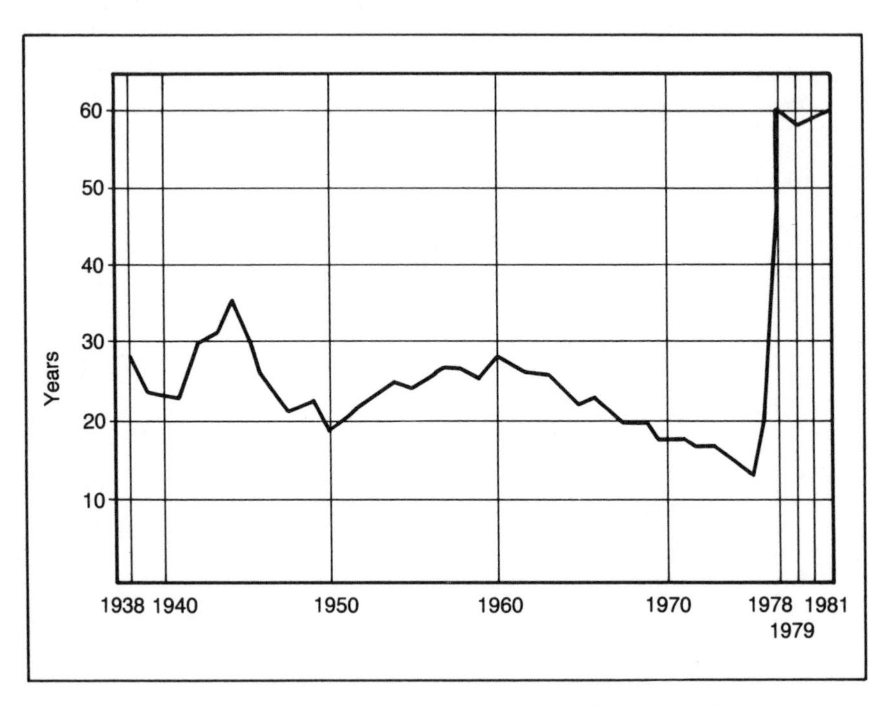

Fig. D.2 *Reserves/production ratio's dramatic climb (source: Pemex)*

Appendix E

Mexican exploration/drilling

Table E.1
How drilling looks in Mexico

	1976	1977	1978	1979	1980	1981	Percent variation 1976–81 Total	1976–81 Average Annual	1981–80 Annual
Average rigs in operation	131	155	176	193	200	205	56	9	3
Development	78	95	107	123	135	135	73	12	0
Exploration	53	60	69	70	65	70	32	6	8
Exploration and development									
Wells completed	336	307	307	333	434	412	23	4	−5
Gas	54	62	54	52	60	81	50	8	35
Oil	171	144	144	167	231	213	25	4	−8
Water and gas injection	—	—	4	14	35	27	—	—	−23
Unproductive	111	101	105	100	108	91	−18	−4	−16
Thousand feet drilled	3,078	3,006	3,001	3,200	4,105	4,138	34	6	1

Source: Secretariado Técnico de la Comisión de Energéticos

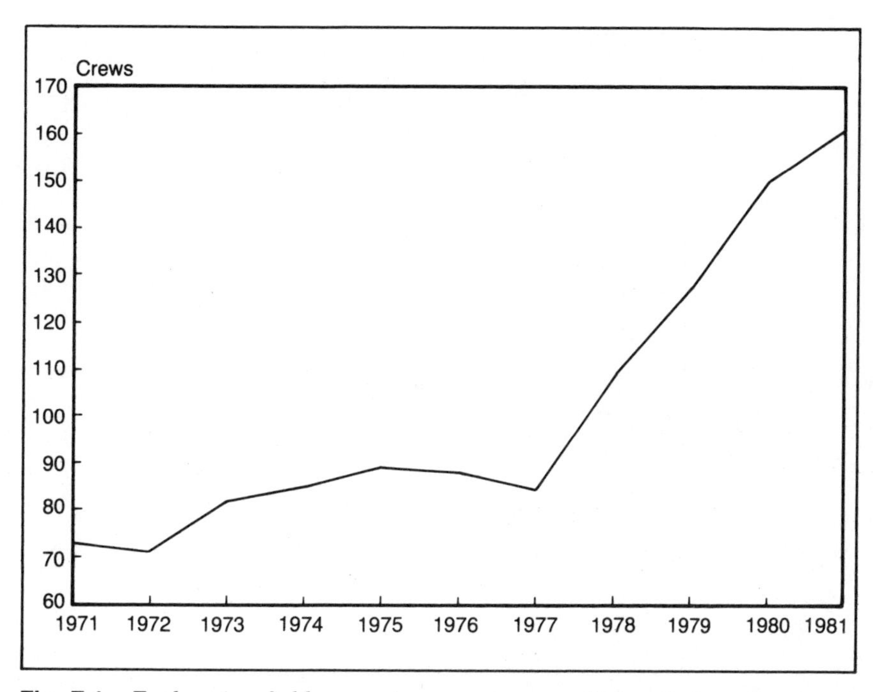

Fig. E.1 *Exploration field crews increase (source: Baker & Associates)*

Table E.2
A look at completed exploratory wells

Year	*Productive*		*Unproductive**		*Total wells*		*Miles drilled*		*Average depth/well (ft)*
	P	*C*	*P*	*C*	*P*	*C*	*P*	*C*	
1938	3		2		5		8		3,281
1971	26	5	78	11	104	16	647	93	12,739
1972	27	3	96	13	123	16	716	85	11,755
1973	24		57	13	81	13	480	61	11,726
1974	19	1	71	2	90	3	463	13	10,443
1975	12	1	57	4	69	5	373	26	10,994
1976	24	1	49	1	73	2	378	10	10,541
1977	29	1	34	3	63	4	359	21	11,555
1978	25	4	36	7	61	11	344	69	11,709
1979	21	9	30	7	51	16	259	106	11,115
1980	29	6	27	7	56	13	235	90	9,606
1981	23	1	42	4	65	5	330	32	10,544

P = Pemex; C = Contractors
*Includes wells capped by mechanical accident to 1963.
Source: Pemex

Table E.3
A look at completed development wells

Year	*Productive**		*Unproductive***		*Total wells*		*Miles drilled*		*Average depth/well (ft)*
	P	*C*	*P*	*C*	*P*	*C*	*P*	*C*	
1938	8		4		12		19		3,281
1971	311	5	66	1	379	6	1,427	35	7,785
1972	222	8	53	2	275	10	1,104	50	8,254
1973	243	6	61	2	304	8	1,164	35	7,835
1974	230	14	50	7	280	21	1,151	89	8,392
1975	206	6	47	3	253	9	1,048	40	8,464
1976	194	6	48	4	242	10	1,032	50	8,750
1977	168	8	43	1	211	9	993	35	9,531
1978	173		37	1	210	1	933	14	9,157
1979	186	17	37		223	17	998	100	9,324
1980	252	39	35	5	287	44	1,228	298	9,396
1981	252	45	35	10	287	55	1,275	393	9,938

P = Pemex; C = Contractors
*Includes water injection
**Includes wells capped by mechanical accident, to 1963
Source: Pemex

Table E.4
A look at total wells completed

Year	Wells		Miles drilled		Average depth/well (ft)	
	P	C	P	C	P	C
1938	17		27		3,281	
1971	481	22	2,074	129	8,793	11,929
1972	398	26	1,820	135	9,324	10,600
1973	385	21	1,643	97	8,701	9,373
1974	370	24	1,614	101	8,546	8,612
1975	322	14	1,421	66	8,996	9,609
1976	315	12	1,410	60	9,124	10,115
1977	274	13	1,352	56	10,059	8,832
1978	271	12	1,278	84	9,613	14,216
1979	274	33	1,257	206	9,350	12,726
1980	343	57	1,463	388	8,694	13,871
1981	352	60	1,604	425	9,291	14,436

P = Pemex; C = Contractors
Source: Pemex

Table E.5
Pemex well productivity*

Year	Wells	Total production (million bbl)	Total production/well (b/d)
1938	1,053	44	114
1971	4,455	298	183
1972	4,375	309	193
1973	4,339	318	201
1974	4,043	379	257
1975	4,074	439	295
1976	3,802	469	337
1977	4,079	533	358
1978	4,309	658	418
1979	4,390	785	490
1980	4,706	1,015	589
1981	4,621	1,199	711

*Includes crude, condensate, gas liquids, and dry-gas crude equivalent.
Source: Pemex

Table E.6
A look at producing fields

Year	Fields discovered: Crude oil	Natural gas	Total	Fields in production
1971	8	9	17	173
1972	11	5	16	179
1973	6	6	12	169
1974	4	4	8	189
1975	3	3	6	195
1976	8	11	19	230
1977	17	9	26	206
1978	11	9	20	256
1979	17	8	25	298
1980	28	6	34	323
1981	10	11	21	323

Source: Pemex

Table E.7
Number of exploratory wells constant in 1982, but success rate drops

District	Productive Gas	Productive Oil	Unproductive	Stratigraphic	Total	Success %	Feet Drilled
Reynosa	4	0	6	0	10	40	107,574
Monclova	3	0	22	0	25	12	269,843
Ebano	0	0	1	2	3	0	46,282
Cerra Azul	0	0	1	0	1	0	6,670
Poza Rica	0	5	1	2	8	83	63,569
Papaloapan	0	0	1	0	1	0	27,710
El Plan	0	0	1	1	2	0	39,583
Nanchital	0	0	0	0	0	—	0
Agua Dulce	0	0	1	0	1	0	10,902
Comalcalco	1	3	9	0	13	31	263,432
							0
Ciudad Pemex	0	0	3	0	3	0	62,253
Gulf of Campeche	0	2	1	0	3	67	45,334
							0
Total 1982	8	10	47	5	70	28	943,151
Total 1981	12	12	39	7	70	38	740,053
Average depth per well			1982: 13,474 ft				
			1981: 10,571 ft				

Note: the success rates exclude stratigraphic wells
Source: Pemex

Appendix F
Mexico's hydrocarbon production

Table F.1
Where crude oil production goes

	(1,000 b/d)						Percent variation		
							1976–81	*1976–81*	*1981–80*
	1976	*1977*	*1978*	*1979*	*1980*	*1981*	*Total*	*Average*	*Annual*
Total production	801	981	1,213	1,471	1,936	2,313	189	24	19
Inventory variation	2	0	4	−14	−10	4	124	17	−140
Total supply	803	981	1,216	1,457	1,926	2,317	189	24	20
Exports	94	202	365	533	828	1,098	1,063	63	33
Total internal supply available	708	779	851	925	1,099	1,219	72	11	11
Pemex refineries	666	751	790	849	996	1,107	66	11	11
Crude for product exchange	26				7	15	−43	−11	124
Losses and statistical discrepancies	16	28	62	77	96	97	502	43	1

Source: Secretariado Técnico de la Comisión de Energéticos

Table F.2
Important crude oil fields in the Southern Zone

	(1,000 b/d)						Percent variation		
							1976–81	*1976–81*	*1981–80*
	1976	*1977*	*1978*	*1979*	*1980*	*1981*	*Total*	*Average*	*Annual*
Southern Zone total	604.10	786.60	1,008.60	1,274.00	1,752.80	2,125.10	251.78	28.60	21.24
Cretaceous	451.30	647.20	865.40	1,067.50	998.50	914.10	102.55	15.16	−8.45
Sitio grande	36.20	50.70	67.70	87.80	91.60	69.30	91.44	13.87	−24.34
Cactus	77.50	98.00	122.00	117.10	92.60	40.30	−48.00	−12.26	−56.48
Samaria	240.80	291.50	315.60	304.60	301.40	283.50	17.73	3.32	−5.94
Cunduacan	75.30	158.90	190.60	197.30	145.50	99.60	32.27	5.75	−31.55
Oxiacaque	—	1.10	33.40	91.80	62.60	41.00	—	—	−34.50
Nispero	10.80	21.90	35.50	40.30	62.10	53.00	390.74	37.46	−14.65
Iride	10.10	18.00	37.80	59.60	62.30	60.20	496.04	42.91	−3.37
Huimanguillo	—	1.00	30.30	120.20	122.30	201.10	—	—	64.43
Giraldas	0.00	0.00	2.10	12.10	23.60	52.50	—	—	122.46
Jujo	0.00	0.00	0.00	0.60	1.60	15.00	—	—	837.50
Agave	0.00	1.00	13.90	34.50	40.50	53.10	—	—	31.11
Paredon	0.00	0.00	2.30	20.60	23.80	33.10	—	—	39.08
Chiapas	0.00	0.00	0.00	0.40	4.50	13.30	—	—	195.56
Other	0.00	0.00	12.10	52.10	28.30	34.00	—	—	20.14
Other Cretaceous	—	5.90	32.50	48.80	58.10	65.40	—	—	12.56
Tertiary	152.8	139.40	143.10	154.80	140.90	128.50	—	—	−8.80
Campeche	—	—	—	51.80	613.40	1,082.50	—	—	76.48
Cantarell	—	—	—	51.80	610.90	886.60	—	—	45.13
Abkatun	—	—	—	—	2.50	119.80	—	—	4,692.00
Others	—	—	—	—	—	76.20	—	—	—

Source: Secretariado Técnico de la Comisión de Energéticos

Table F.3
Important associated gas fields in the Southern Zone

	(MMcfd)						Percent variation		
							1976–81	*1976–81*	*1981–80*
	1976	*1977*	*1978*	*1979*	*1980*	*1981*	*Total*	*Average*	*Annual*
Southern Zone total	831	975	1,355	1,923	2,311	2,792	236	27	21
Cretaceous	633	821	1,205	1,747	1,915	2,132	237	27	11
Sitio Grande	555	81	106	123	112	95	−83	−30	−15
Cactus	135	178	215	221	133	97	−28	−6	−27
Samaria	286	328	354	375	335	318	11	2	−5
Cunduacan	99	163	170	203	208	143	45	8	−31
Oxiacaque	0	2	29	111	165	143	—	—	−13
Nispero	18	36	54	62	89	81	359	36	−9
Iride	15	24	37	57	59	84	449	41	44
Huimanguillo area	0	2	158	397	603	997	—	—	65
Giraldas	0	0	10	53	101	237	—	—	134
Jujo	0	0	0	0	2	19	—	—	1,133
Agave	0	2	108	254	296	457	—	—	54
Paredon	0	0	5	43	71	84	—	—	18
Chiapas	0	0	0	2	21	38	—	—	80
Others	0	0	35	46	112	162	—	—	45
Other Cretaceous	25	7	85	198	213	174	589	147	−18
Tertiary onshore	198	154	150	153	137	130	−34	92	−5
Campeche	0	0	0	23	258	530	—	—	105
Cantarell	0	0	0	23	257	383	—	—	49
Abkatun	0	0	0	0	1	103	—	—	7,279
Others	0	0	0	0	0	43	—	—	—

Source: Secretariado Técnico de la Comisión de Energéticos

Table F.4
Southern Zone contributes to crude and natural gas production

*(crude = 1.000 bbl)**
(gas = MMcf)
District

	Istmo		*Tabasco*		*Agua Dulce*		*Cuidad Pemex*	
Year	*Crude oil*	*Gas*	*Crude oil*	*Gas*	*Crude oil*	*Gas*	*Crude oil*	*Gas*
1971	17,952	23,943	57,783	274,925				
1972					39,100	50,465	189	232,512
1973					37,112	47,180	241	245,437
1974					34,730	44,497	220	252,359
1975					33,117	38,811	251	265,425
1976					26,783	31,183	205	237,103
1977					23,867	27,086	140	161,883
1978					22,957	26,910	82	143,448
1979					21,612	25,250	40	45,485
1980					20,571	23,166	47	157,115
1981					18,442	21,401	28	184,802

*Includes crude and condensate.
Source: Pemex

Table F.5
Southern Zone contributes to crude and natural gas production

(crude = 1.000 bbl)*
(gas = MMcf)
District

Year	Comalcalco		El Plan		Nanchital		Gulf of Campeche		Total	
	Crude oil	*Gas*	*Crude oil*	*Gas*	*Crude oil*	*Gas*	*Crude oil*	*Gas*	*Crude oil*	*Gas*
1971									75,735	298,868
1972	22,550	23,626	19,988	32,631	1,977	918			83,804	340,151
1973	33,556	41,636	16,443	30,547	1,898	1,059			89,250	365,860
1974	81,007	125,897	13,702	28,711	2,063	1,695			131,722	453,158
1975	135,169	203,519	12,613	28,676	2,396	1,624			183,546	538,055
1976	180,039	247,662	11,518	22,284	2,554	1,519			221,099	539,750
1977	249,255	312,288	11,512	14,479	2,423	1,307			287,197	517,043
1978	328,572	451,322	15,534	14,867	2,198	1,448			369,343	637,995
1979	402,858	647,283	23,170	19,529	2,008	1,236	18,887	8,334	468,575	747,118
1980	373,314	708,307	21,192	18,258	1,893	1,342	224,511	94,502	641,528	1,002,690
1981	341,409	785,081	18,912	19,352	1,760	1,624	395,114	193,242	775,665	1,205,503

*Includes crude and condensate.
Source: Pemex

Table F.6
Production growth for crude oil and natural gas

Year	*Crude oil (1,000 bbl)**	*Natural gas (MMcf)*
1938	38,482	24,085
1971	155,911	643,434
1972	161,367	660,244
1973	164,909	676,771
1974	209,855	744,681
1975	261,589	786,494
1976	293,117	771,803
1977	358,090	746,871
1978	442,607	934,921
1979	536,926	1,064,597
1980	708,593	1,298,592
1981	844,241	1,482,229

*Includes crude and condensate.
Source: Pemex

Table F.7
Crude production drops in all zones in 1982, except in the Gulf of Campeche

Zone	*Barrels* Annual total	*Average/day*
Northern		
N.W. Frontera district	150,015	411
North district	11,753,730	32,202
South district	7,039,025	19,285
Total 1982	18,942,770	51,898
1981	19,764,750	54,150
Variation		−4.2%
Central		
Poza Rica	42,287,075	115,855
Papaloapan Basin	4,278,530	11,722
Total 1982	46,565,605	127,577
1981	48,502,660	132,884
Variation		−4.0%
Southern		
Agua Dulce district	17,536,245	48,045
El Plan district	16,261,115	44,551
Nanchital district	1,526,430	4,182
Total 1982	35,323,970	96,778
1981	39,113,765	107,161
Variation		−9.7%
Southeastern		
Comalcalco district		
Tertiary	6,704,320	18,368
Mesozoic	304,533,370	834,338
Ciudad Pemex district	6,570	18
Total 1982	311,244,260	852,724
1981	341,437,060	935,444
Variation		−8.8%
Offshore		
Gulf of Campeche		
Total 1982	590,353,190	1,617,406
1981	395,114,325	1,082,505
Variation		+49.4%
Total crude 1982	1,002,429,795	2,746,383
1981	843,932,560	2,312,144
Variation		+18.8%

Note: In Pemex's annual report for 1982, *Memoria de labores,* Pemex has adopted new categories for reporting crude production by geographic region. The three zones in this table, Southern, Southeastern, and Offshore, were reported in earlier *Memorias* as one region, the Southern Zone, and have been cited under that heading in the text of this book.
Source: Pemex

Table F.8
Crude oil production high in spring and autumn

Average production	*(1,000 b/d)*		*January = 100*
	1981	*1982*	*by month*
January	2.219	2.313	100.00
February	2.120	2.551	112.58
March	2.366	2.542	108.30
April	2.538	2.778	117.30
May	2.545	2.712	116.00
June	2.558	2.788	117.96
July	2.096	2.786	107.72
August	2.261	2.791	111.47
September	2.477	2.828	117.06
October	2.488	2.898	118.84
November	2.090	2.937	129.61
December	1.981	3,023	110.41
Annual	2.3529	2.7456*	

*The official figure is 2.746383 b/d for 1982
Source: American Chamber of Commerce of Mexico (production figures)

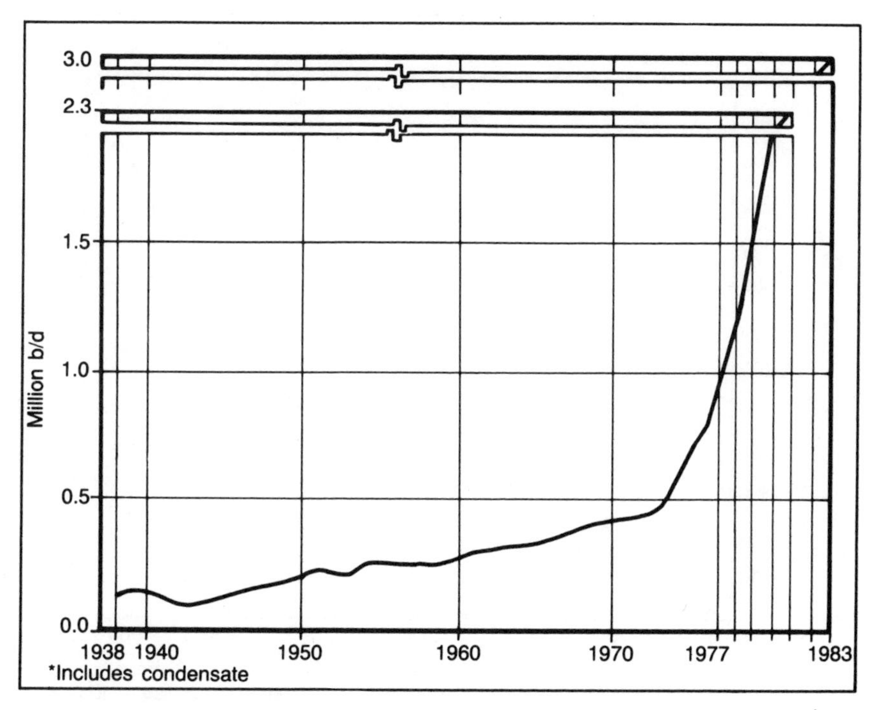

Fig. F.1 *Crude oil production's steady climb* (source: Pemex)*

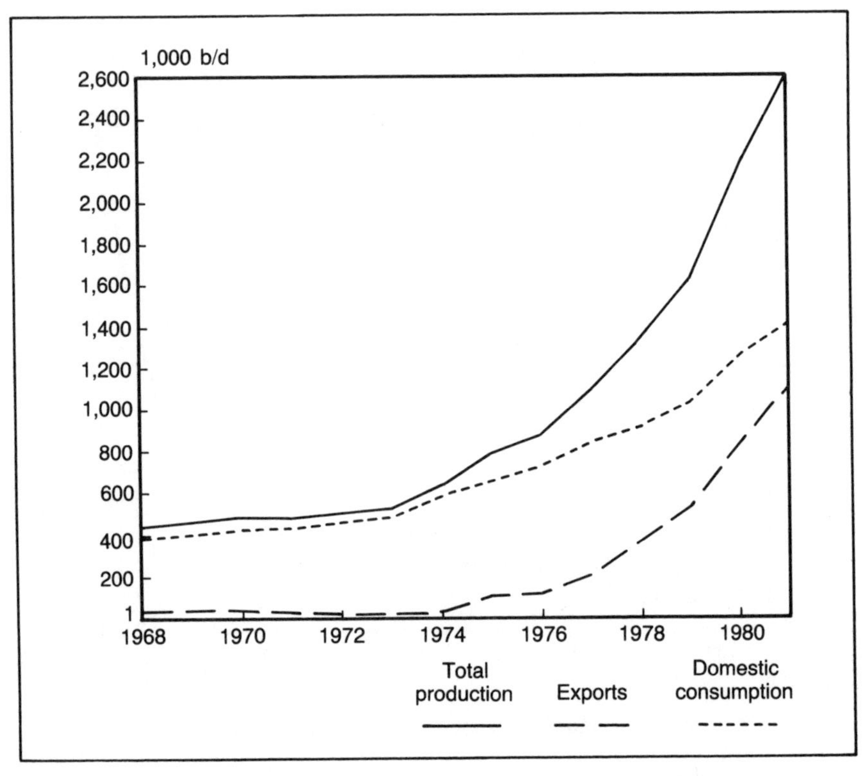

Fig. F.2 *How Mexico exploited crude oil (source: Diemex-Wharton)*

Appendix G

Pemex refining

Table G.1
Refinery runs and output on upswing

	(1,000 b/d)						Percent variation		
	1976	*1977*	*1978*	*1979*	*1980*	*1981*	*1976–81 Total*	*1976–81 Average*	*1981–80 Annual*
Total runs	740	836	882	965	1,149	1,272	72	11	11
Crude	666	751	790	847	996	1,107	66	11	11
Gas liquids	74	85	91	118	153	165	123	17	8
Reprocessed	9	10	13	17	12	19	116	17	51
Total production	733	824	877	959	1,140	1,260	72	11	11
Dry gas	14	20	20	23	29	30	109	16	2
LPG	56	66	74	91	120	136	144	20	13
Gasoline	212	213	246	285	328	361	70	11	10
Kerosene	54	55	58	65	69	70	31	6	2
Diesel	163	182	199	215	244	270	66	11	11
Residual fuel	200	233	244	238	309	347	74	12	12
Others	25	23	25	27	30	33	34	6	11
Delivery to petrochemical operations	10	12	11	15	12	14	33	6	17
Gravity of crude processed	31.4	31.3	31.2	31.1	29.6	29.5			

Source: Secretariado Técnico de la Comisión de Energéticos

Table G.2
How Pemex measures nominal refining capacity

	(b/d)			
Year	*Primary distillation*	*Index 1938 = 100*	*Catalytic/thermal crackers/visbreakers*	*Index (1938 = 100)*
1938	102,000	100.00	12,000	100.00
1971	592,000	580.39	150,000	1,250.00
1972	625,000	612.75	168,000	1,400.00
1973	760,000	745.10	168,000	1,400.00
1974	760,000	745.10	168,000	1,400.00
1975	785,000	769.61	168,000	1,400.00
1976	968,500	949.51	208,000	1,733.33
1977	973,500	954.41	277,000	2,308.33
1978	988,500	969.12	317,000	2,641.67
1979	1,341,000	1,314.71	317,000	2,641.67
1980	1,476,000	1,447.06	397,000	3,308.33
1981	1,523,500	1,493.63	397,000	3,308.23

Source: Pemex

Table G.3
What refineries yielded

(percent of volume)

Year	*Liquid gas*	*Gasoline*	*Jet fuel*	*Kerosenes*	*Diesel*
1971	7.43	28.63	1.78	6.29	16.43
1972	7.47	28.66	1.84	5.87	17.49
1973	8.15	27.99	2.11	5.87	18.16
1974	7.83	27.89	2.38	5.36	21.32
1975	7.61	27.27	2.25	5.13	22.24
1976	7.45	28.34	2.26	4.92	21.74
1977	7.28	27.35	2.51	4.02	21.56
1978	8.27	27.49	2.26	4.24	22.18
1979	9.23	29.04	2.56	4.10	21.93
1980	10.31	28.23	2.37	3.57	21.03
1981	10.53	27.96	2.24	3.19	20.92

Year	*Residual fuel*	*Asphalt*	*Lubricants*	*Wax*
1971	25.01	4.86	1.06	0.28
1972	26.43	4.38	1.12	0.28
1973	25.97	3.65	1.09	0.29
1974	26.44	2.01	1.14	0.28
1975	26.31	1.66	1.19	0.23
1976	26.73	1.57	1.12	0.22
1977	27.57	1.43	0.88	0.22
1978	27.24	1.48	0.87	0.23
1979	24.20	1.50	0.77	0.18
1980	26.57	1.45	0.66	0.19
1981	26.89	1.41	0.73	0.14

Source: Pemex

Table G.4
Gradual climb in refinery output

	(1,000 b/d)					
	1976	*1977*	*1978*	*1979*	*1980*	*1981*
Total*	733.1	823.7	877.3	958.7	1,139.7	1,260.2
Refinery:						
Azcapotzalco	106.7	96.6	100.8	99.3	100.2	88.1
Cadereyta	—	—	—	64.9	151.4	194.1
Madero	171.8	148.6	170.8	149.7	163.2	160.6
Minatitlan	239.9	250.2	254.9	279.6	258.2	258.3
Poza Rica	28.1	25.8	27.4	23.6	28.2	28.0
Reynosa	11.3	9.2	10.7	10.7	9.7	8.5
Salamanca	141.2	164.9	159.5	159.5	160.7	177.0
Salina Cruz	—	—	—	16.8	118.1	135.0
Tula	34.1	128.4	153.2	144.6	131.8	128.2
Cactus**	—	—	—	10.0	18.2	82.4

*Includes the production of petroleum products that result from the processing of LPG.
**Preliminary
Source: Secretariado Técnico de la Comisión de Energéticos

Table G.5
Primary distillation capacity rises

	(1,000 b/d)					
	1976	*1977*	*1978*	*1979*	*1980*	*1981*
Primary distillation	865	865	865	1,135	1,270	1,270
Refinery:						
Azcapotzalco	105	105	105	105	105	105
Cadereyta	—	—	—	100	235	235
Madero	185	185	185	185	185	185
Minatitlan	200	200	200	200	200	200
Poza Rica	16	16	16	16	16	16
Reynosa	9	9	9	9	9	6
Salamanca	200	200	200	200	200	200
Salina Cruz	—	—	—	170	170	170
Tula	150	150	150	150	150	150
Gas liquids	104	109	125	207	207	254
Processing plant:						
Cactus	—	—	—	83	83	83
Cangrejera	—	—	—	—	—	47
Madero	—	—	1	1	1	1
Minatitlan	70	75	90	90	90	90
Poza Rica	22	22	22	22	22	22
Reynosa	12	12	12	12	12	11
Total capacity	969	974	990	1,342	1,477	1,524

Source: Secretariado Técnico de la Comisión de Energéticos

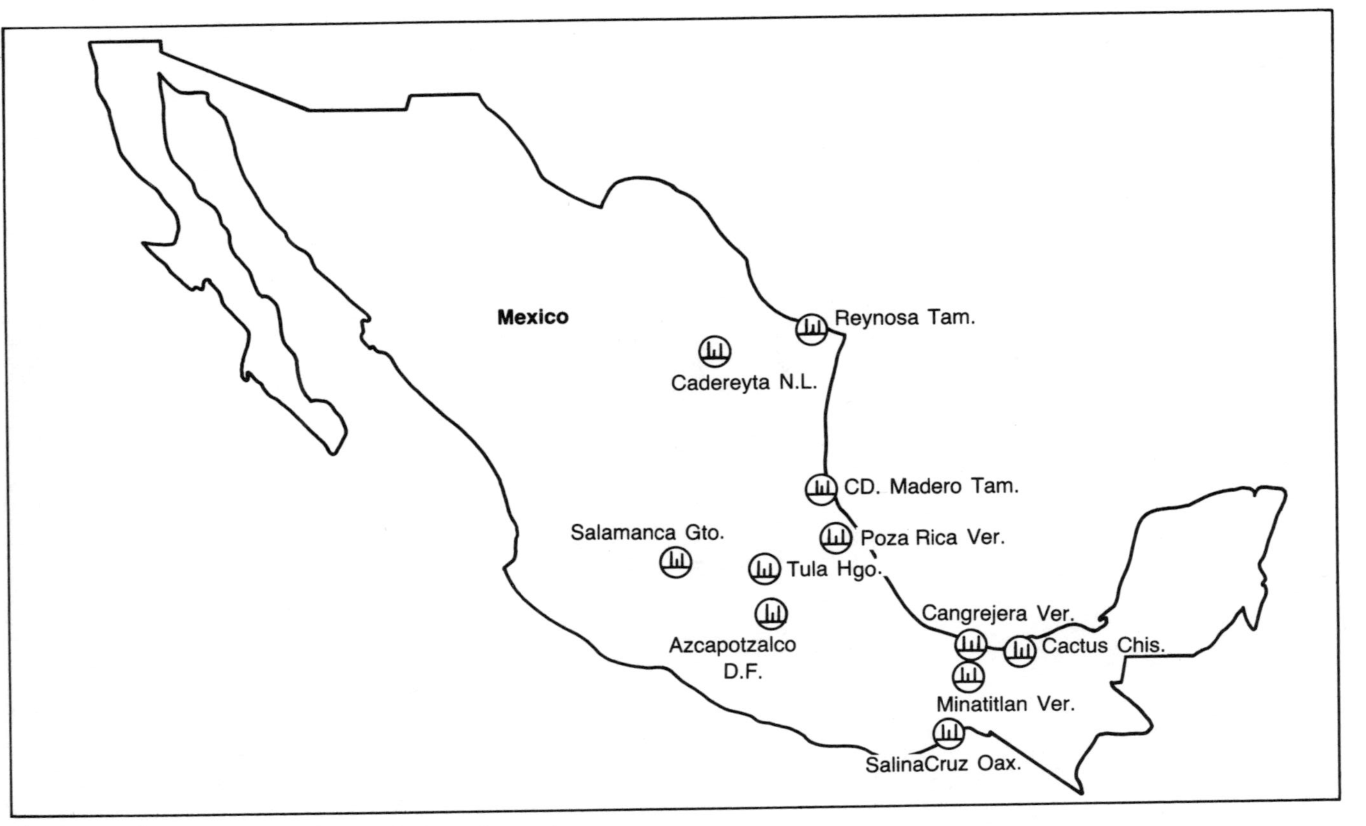

Fig. G.1 *Refining centers 1981 (source: Pemex)*

Appendix H
Mexican petrochemicals

Table H.1
Wonder industry's production surges

(1,000 tons)

	1976	*1977*	*1978*	*1979*	*1980*	*1981*	*1976–81 Total*	*1976–81 Average*	*1981–80 Annual*
Total	4,037	4,297	5,922	6,492	7,392	9,371	132	18	27
Acetaldehyde	48	45	46	50	49	125	162	21	157
Acrylonitrile	23	20	20	24	56	55	145	20	−1
Ammonia	885	966	1,615	1,691	1,927	2,234	152	20	16
Benzene	101	76	81	72	81	78	−23	−5	−4
Butadiene	19	24	19	18	17	13	−34	−8	−28
Cyclohexane	44	39	38	30	41	45	2	0.41	11
Vinyl chloride	62	57	57	57	64	58	−5	−1	−9
Ethylene dichloride	0	100	99	81	109	120	—	—	10
Dodecylbenzene	65	64	64	64	57	61	−6	−1	7
Ethylene	233	235	264	351	374	387	66	11	4
Ethyl oxide	25	27	27	25	30	50	98	15	63
Hexane	31	31	30	55	61	67	118	17	9
Methanol	33	34	105	178	177	184	458	41	4
Propylene	116	141	14,123	164	140	160	38	7	14
Styrene	36	37	39	33	32	34	−6	−1	5
Toluene	135	119	132	110	128	135	−1	−0.11	5
Xylene	128	115	133	121	136	145	14	3	7
Others	1,983	2,129	2,910	3,123	3,478	3,949	99	15	14
Low-density polyethylene	96	97	99	98	94	93	−3	−1	0
High-density polyethylene	0	0	3	60	68	80	—	—	17

Source: Secretariado Técnico de la Comisión de Energéticos

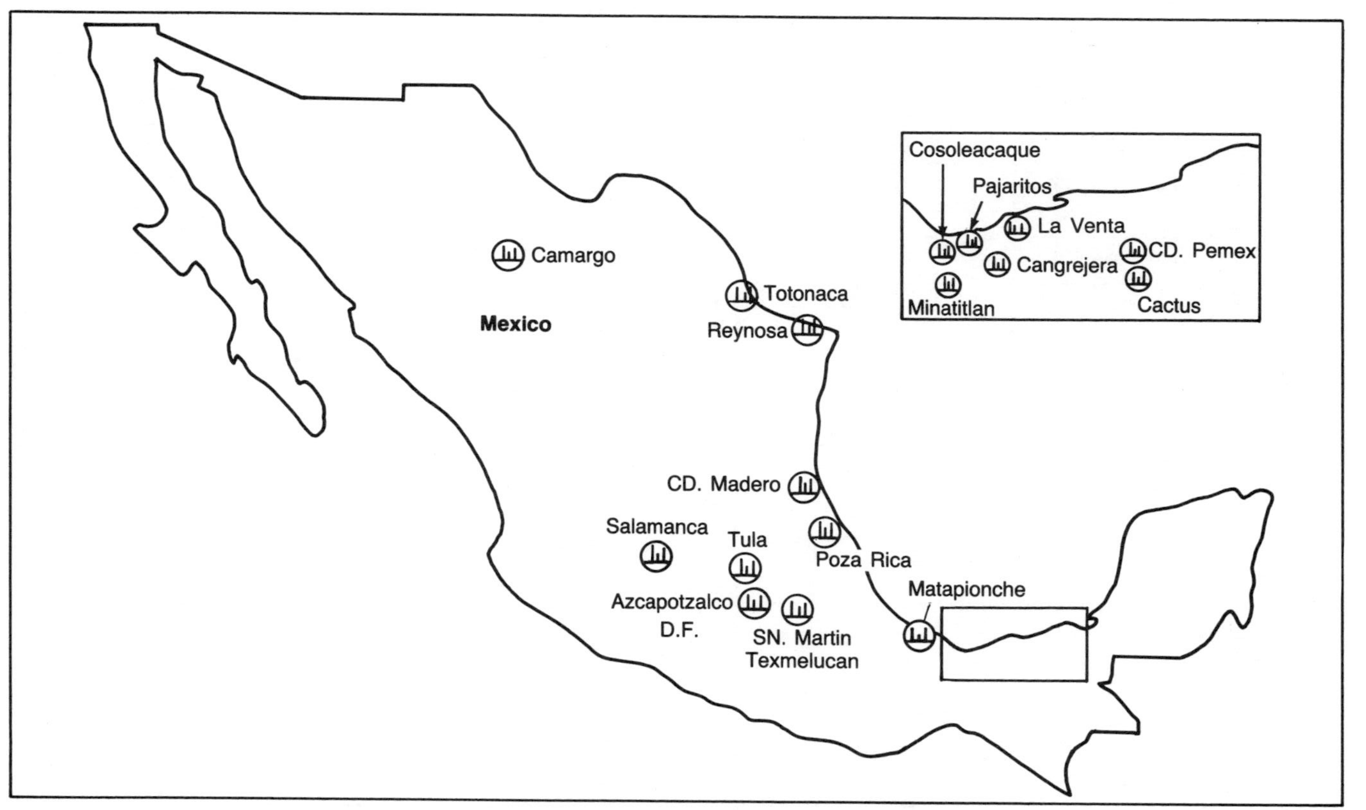

Fig. H.1 *Petrochemical centers 1981 (source: Pemex)*

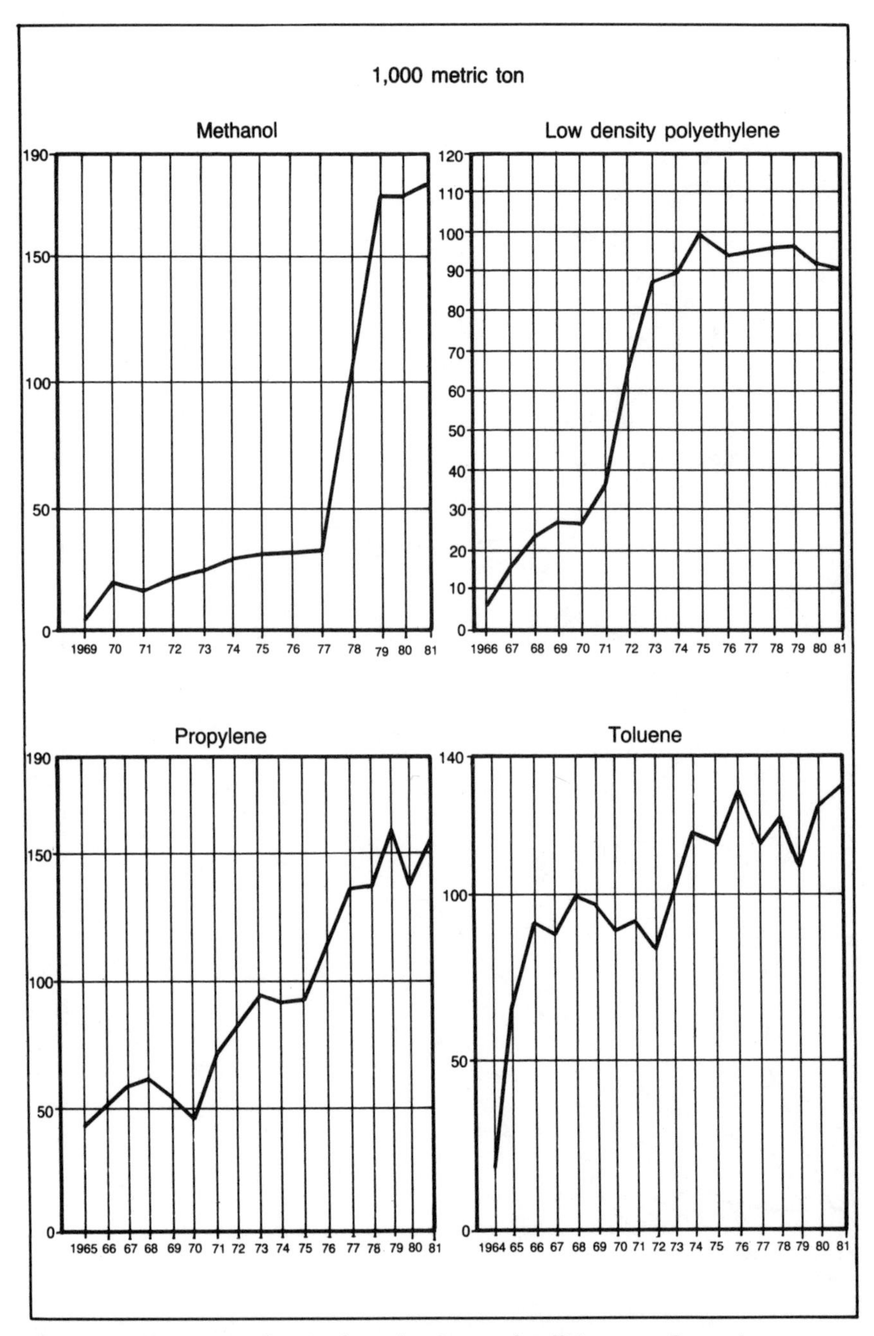

Fig. H.2 *How petrochemical production took off (source: Pemex)*

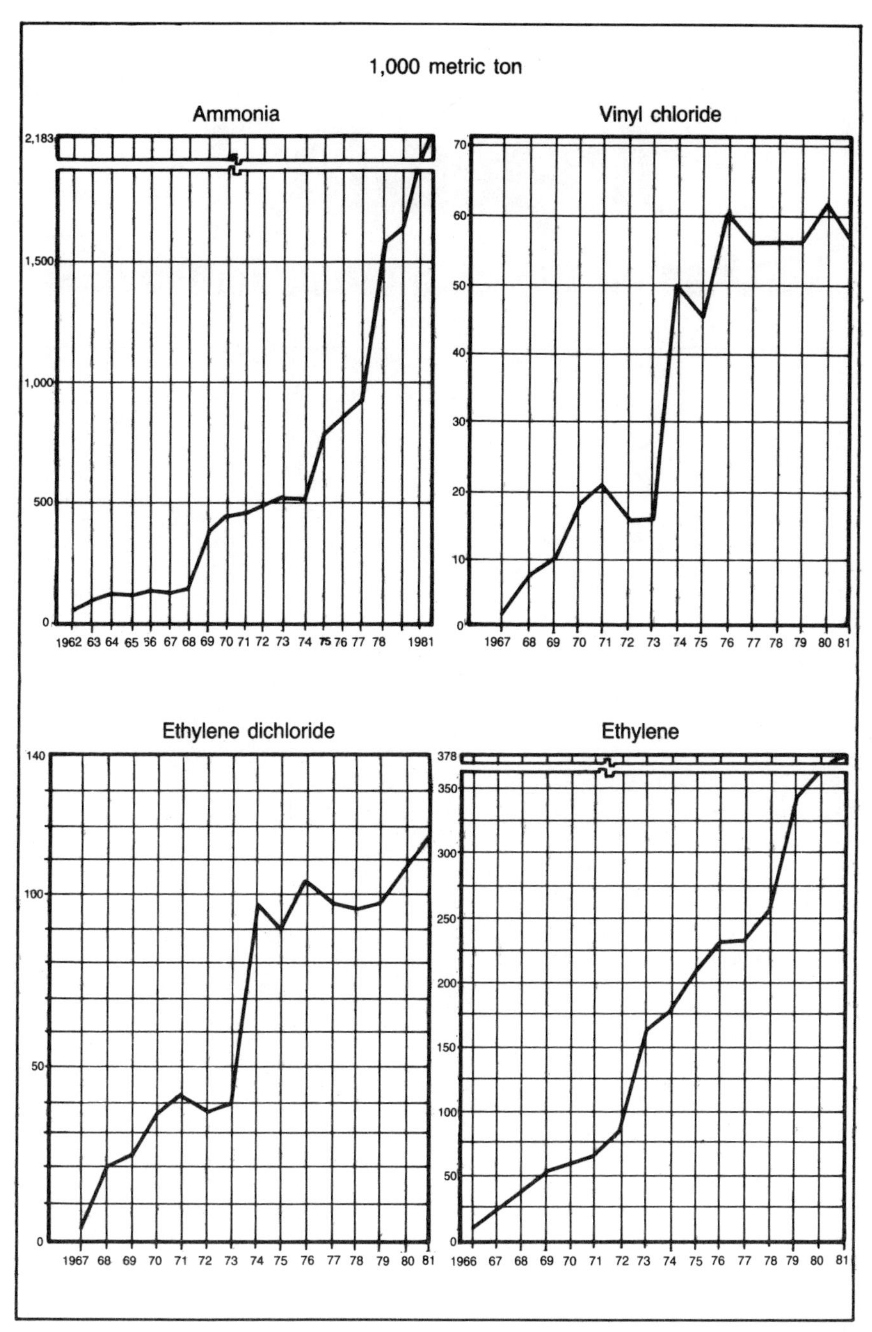

Fig. H.2 *How petrochemical production took off (continued), (source: Pemex)*

Appendix I
Mexican oil transportation/distribution

Table I.1
Where Pemex has maritime and land transport capacity

Year	Oil tankers	(1,000 bbl) Truck tankers	Rail tankers
1972	2,709	162	803
1973	2,883	262	861
1974	3,460	327	916
1975	3,638	347	987
1976	4,055	422	1,130
1977	4,903	535	1,223
1978	5,117	631	1,232
1979	6,222	820	1,252
1980	6,804	1,007	1,219
1981	6,790	1,189	1,071

Source: Pemex

Table I.2
Where Pemex concession holders operate

State	1976	1977	1978	1979	1980	1981
	(Number of service stations)					
Aguascalientes	18	19	19	21	22	23
Baja California Norte	268	268	261	257	260	260
Baja California Sur	34	35	34	34	34	35
Campeche	17	18	18	17	17	18
Coahuila	87	80	81	85	85	87
Colima	16	15	16	15	16	17
Chiapas	61	61	64	64	64	66
Chihuahua	200	204	204	203	201	204
Federal District	241	247	248	248	248	255
Durango	59	59	59	61	61	62
Guanajuato	85	88	90	92	96	98
Guerrero	58	56	58	57	61	62
Hidalgo	62	61	67	67	70	71
Jalisco	171	172	173	177	181	186
México	148	153	154	160	168	178
Michoacán	112	117	119	121	119	121
Morelos	29	30	29	30	32	32
Nayarit	35	35	36	34	33	34
Nuevo León	100	105	107	107	112	115
Oaxaca	59	63	66	56	64	65
Puebla	105	108	108	109	108	111
Querétaro	22	24	24	23	26	26
Quintana Roo	12	12	14	15	14	15
San Luis Potosí	55	55	55	61	60	62
Sinaloa	83	86	86	88	89	91
Sonora	167	154	156	158	165	168
Tabasco	25	26	26	26	29	30
Tamaulipas	104	104	105	111	113	118
Tlaxcala	24	25	25	24	26	26
Veracruz	148	149	151	141	150	154
Yucatán	48	52	54	54	56	56
Zacatecas	55	58	57	57	58	60
Total	2,708	2,739	2,764	2,773	2,838	2,906

Source: Pemex

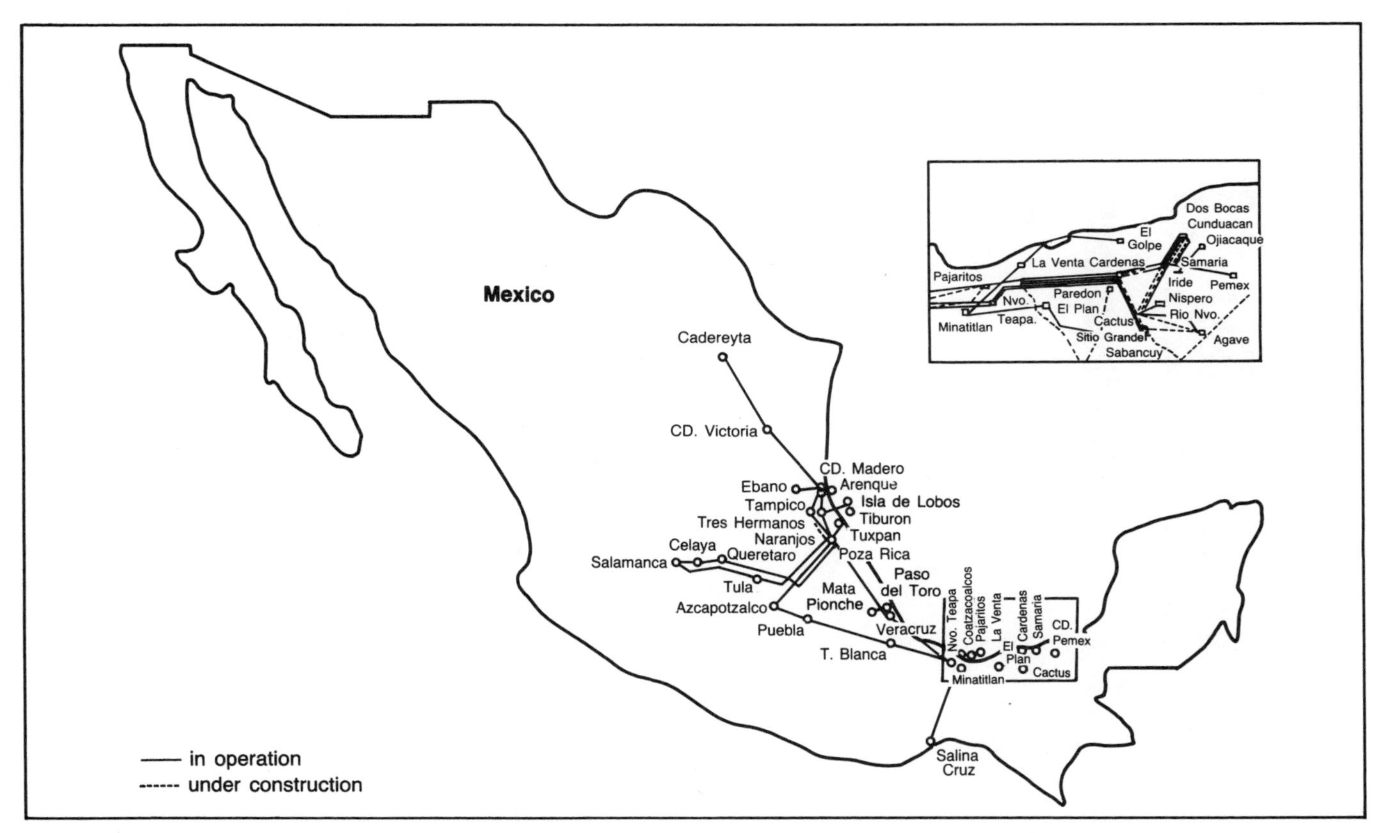

Fig. I.1 *Oil pipelines 1981 (source: Pemex)*

Appendix J

Mexico's petroleum sales

Table J.1
Who was lifting Pemex crude in 1980

(1,000 b/d)

Volume level	*Buyer*	*Volume*
50–150	Exxon	141
	Shell Oil Co.	135
	Spain	95
	Ashland	80
	Israel	53
20–50	France	39
	Clark Refining	37
	Amoco	34
	Japan	30
	Coastal States Refining	26
0.3–20	Arco	17
	Mobil	17
	Cities Service	17
	Tenneco	17
	Brazil	15
	Dow Chemical	14
	Charter Oil Co.	12
	Texaco	12
	Union	6
	Phillips	6
	Costa Rica	5
	Crown	2
	Sun	2
	Others	4

Source: Pemex

Table J.2
How selected products fared in trade

	(1,000 b/d)						Percent variation		
							1976–81	*1976–81*	*1981–80*
	1976	*1977*	*1978*	*1979*	*1980*	*1981*	*Total*	*Average*	*Annual*
LPG									
Production	56	66	74	91	120	136	144	20	13
Imports	10	3	6	8	8	5	−50	−13	−37
Exports	—	—	—	6	15	3	—	—	−82
Domestic sales	63	62	77	87	103	114	80	12	11
Gasoline									
Production	212	231	246	285	328	361	70	11	10
Imports	7	0	0	0	0	1	−81	−28	1,200
Exports	3	3	2	—	1	3	−4	−1	92
Domestic sales	208	219	238	276	314	358	72	12	14
Kerosene									
Production	54	55	58	65	69	70	31	6	2
Imports	—	0	0	—	1	1	—	—	0
Exports	0.6	1	1	1	0	0	—	—	0
Domestic sales	51	53	55	61	65	67	30	5	2
Diesel									
Production	163	182	199	215	244	270	66	11	11
Imports	3	1	3	—	1	0	−91	−38	−67
Exports	0	1	0	0	1	9	2,733	95	608
Domestic sales	164	169	188	203	215	233	42	7	8
Residuals									
Production	200	233	244	238	309	347	74	12	12
Imports	4	4	19	15	—	—	−100	−100	—
Exports	1	1	—	4	29	52	10,360	153	81
Domestic sales	195	196	229	222	243	250	28	5	3

Source: Secretariado Técnico de la Comisión de Energéticos

Table J.3
Quantity of petroleum products sold in Mexico

	(1,000 bbl)					
Year	*Liquefied gas*	*Gasoline*	*Jet fuel*	*Other middle distillates*	*Diesel*	*Residual fuel*
1971	12,995	55,790	3,303	10,639	28,372	33,158
1972	13,043	60,194	3,802	10,736	33,735	38,779
1973	13,806	66,717	4,441	10,933	36,252	39,845
1974	15,861	68,849	5,897	11,367	44,562	51,785
1975	21,239	70,591	5,751	12,024	55,930	59,883
1976	23,119	75,958	6,336	12,437	59,861	71,412
1977	22,751	80,037	7,034	12,884	61,489	71,513
1978	28,009	86,921	7,347	12,683	68,703	83,622
1979	31,902	100,645	8,974	13,338	74,060	81,150
1980	37,540	114,995	10,052	13,854	78,688	88,972
1981	41,500	130,601	10,522	13,884	85,014	91,104

					Natural gas	
Year	*Lubricants*	*Asphalt*	*Grease*	*Wax*	*(1,000 bbl)**	*(MMcf)*
1971	1,829	3,512	55	522	44,453	301,199
1972	1,963	4,408	58	436	45,532	309,216
1973	2,181	3,880	66	454	49,788	338,138
1974	2,407	3,449	94	520	48,673	330,546
1975	2,524	3,996	79	581	51,961	352,476
1976	2,775	4,087	93	631	49,145	333,371
1977	2,850	3,878	90	714	49,936	338,739
1978	3,332	4,090	92	774	58,817	398,985
1979	3,576	5,084	89	821	72,434	491,333
1980	3,892	5,894	81	794	74,116	502,740
1981	4,121	6,474	91	719	76,164	516,619

*Equivalent to residual fuel.
Source: Pemex

Table J.4
How domestic petrochemical sales increased

	(1,000 tons)					
	1976	*1977*	*1978*	*1979*	*1980*	*1981*
Acetaldehyde	57.09	59.02	65.10	64.08	111.24	136.31
Acrylonitrile	39.75	48.00	56.71	66.30	74.86	81.80
Ammonia	873.06	886.82	880.61	976.37	1,065.69	1,328.06
Benzene	10.70	11.61	11.97	12.14	13.92	14.87
Butadiene	47.90	45.70	57.95	63.22	72.83	76.66
Carbon black feedstock	112.83	114.31	143.84	138.66	142.21	152.47
Carbon dioxide	380.96	422.64	365.15	347.67	454.75	611.00
Cyclohexane	53.11	52.31	46.79	57.31	58.25	57.94
Dodecylbenzene	72.52	90.48	81.63	92.34	99.34	101.95
Ethyloxide	44.98	60.94	66.92	62.39	77.73	85.83
Ethylene dichloride	8.36	10.74	11.97	15.82	14.89	16.90
Heptane	5.29	5.19	3.86	5.21	7.63	7.76
Hexane	32.04	32.17	39.98	42.87	50.86	56.20
Isopropanol	20.67	18.07	20.94	23.55	30.67	34.61
Methanol	74.37	66.88	77.30	93.01	107.86	138.03
Ortho-xylene	23.94	22.97	27.15	31.27	33.89	29.72
Paraxylene	105.16	97.35	96.26	106.65	128.36	205.89
Perfumes	3.97	7.37	7.55	11.21	13.05	13.47
High-density polyethylene	0.00	0.00	0.00	69.57	87.67	107.80
Low-density polyethylene	127.04	144.70	159.50	187.79	247.26	283.85
Styrene	68.72	69.26	82.23	109.40	111.56	128.82
Sulfur	102.89	129.81	147.63	268.07	449.51	471.23
Toluene	63.82	64.73	72.04	84.81	89.64	97.48
Vinyl chloride	81.52	79.75	112.67	132.08	163.29	159.55
Xylene	23.62	21.36	25.73	26.81	28.63	33.44
Others	28.32	24.29	18.02	12.04	15.84	16.70
Total	2,462.66	2,586.56	2,679.49	3,388.02	3,751.43	4,448.29

Source: Pemex

Table J.5
Mexico's preferential crude sales to Caribbean nations

	(1,000 b/d)					
			1982			
	Contracted volume	*1981*	*First quarter*	*Second quarter*	*July 1982*	*Second year of agreement*
Total	76.5	45.8	58.5	50.9	40.5	49.0
Barbados	1.0	—	—	—	—	—
Costa Rica	7.5	5.4	2.5	2.5	6.6	4.2
El Salvador	7.0	5.6	5.8	2.9	8.0	5.6
Guatemala	8.5	5.9	4.2	6.1	3.1	5.1
Honduras	6.0	0.5	—	—	—	—
Jamaica	13.0	7.5	10.4	8.4	—	7.5
Nicaragua	7.5	6.3	5.8	11.4	—	7.1
Panama	12.0	8.2	12.0	11.9	11.7	9.9
Dominican Republic	14.0	6.4	17.8	7.7	11.1	9.6
Haiti	3.5	0.9	—	—	—	—
Value (million $US)		581.9	166.4	140.2	42.2	569.2

Source: *Energéticos,* July 1982, p. 2.

Table J.6
Preferential crude mixes of Mexico's Pact of San José customers
(Isthmus: Maya)

		1982			
	1981	*First quarter*	*Second quarter*	*July 1982*	*Second year of agreement*
Total	73:27	70:30	71:29	75:25	71:29
Barbados	—	—	—	—	—
Costa Rica	100:00	100:00	100:00	100:00	100:00
El Salvador	100:00	100:00	100:00	100:00	100:00
Guatemala	81:19	81:19	77:23	100:00	81:19
Honduras	92:08	—	—	—	—
Jamaica	49:51	43:57	41:59	—	33:67
Nicaragua	100:00	100:00	100:00	—	100:00
Panama	43:57	50:50	51:49	51:49	50:50
Dominican Republic	59:41	73:27	65:35	70:30	63:37
Haiti	40:60	—	—	—	—

Source: *Energéticos,* July 1982, p. 3.

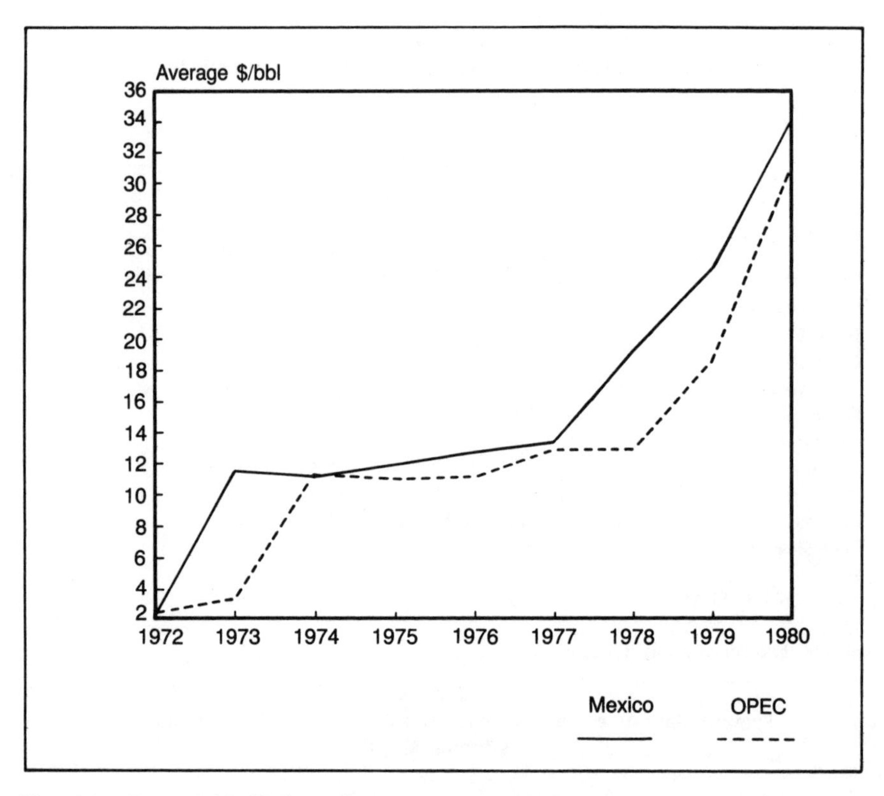

Fig. J.1 *Pemex's bullish crude pricing (source:* Energy Mexico, *1981, p. 37; CIA, "Economic & Energy Indicators," Jan. 7, 1983)*

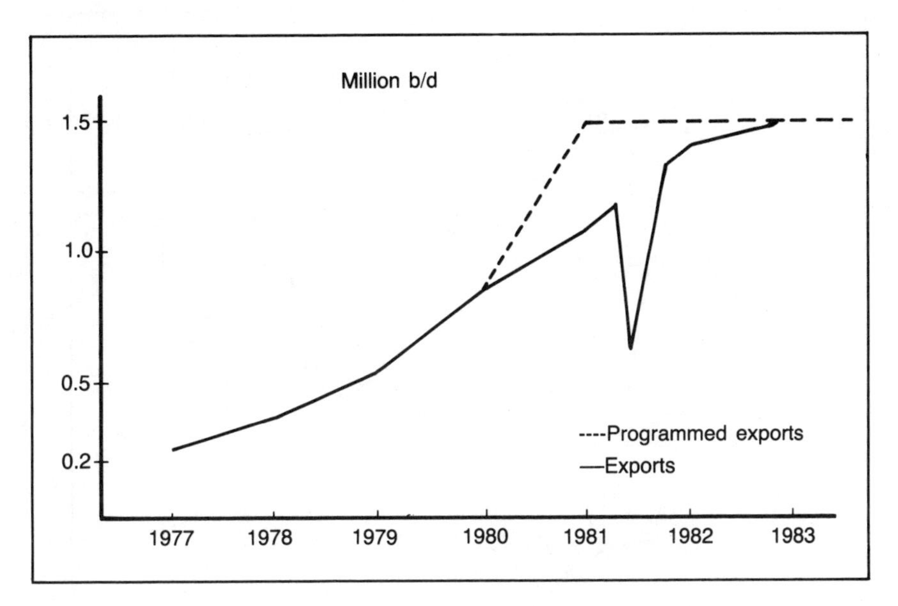

Fig. J.2 *How crude exports compare with planned levels (source: Pemex)*

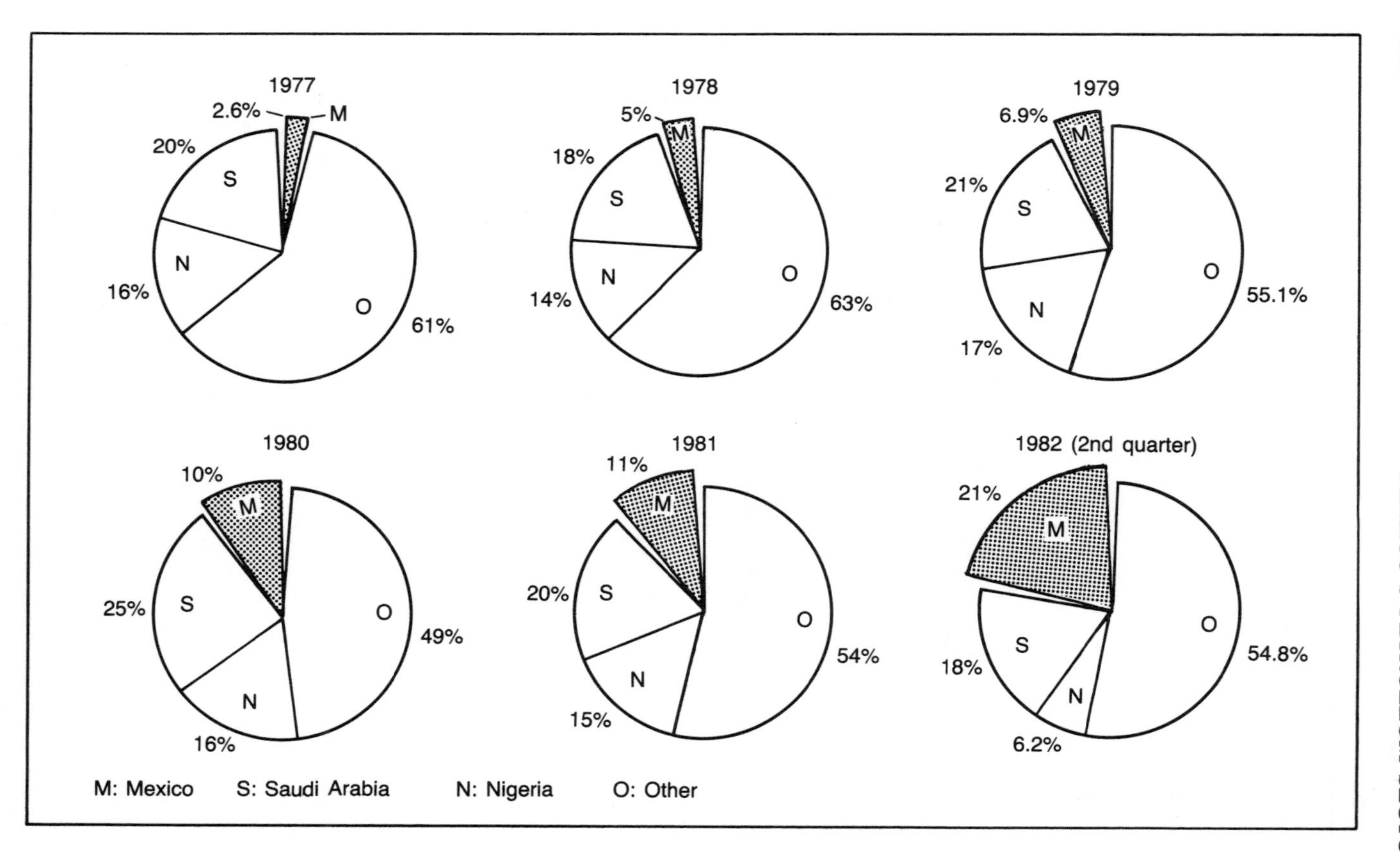
1977
2.6%
M
20%
S
N
16%
O
61%
1978
5%
M
18%
S
N
14%
O
63%
1979
6.9%
M
21%
S
N
17%
O
55.1%
1980
10%
M
25%
S
N
16%
O
49%
1981
11%
M
20%
S
N
15%
O
54%
1982 (2nd quarter)
21%
M
18%
S
N
6.2%
O
54.8%
M: Mexico S: Saudi Arabia N: Nigeria O: Other

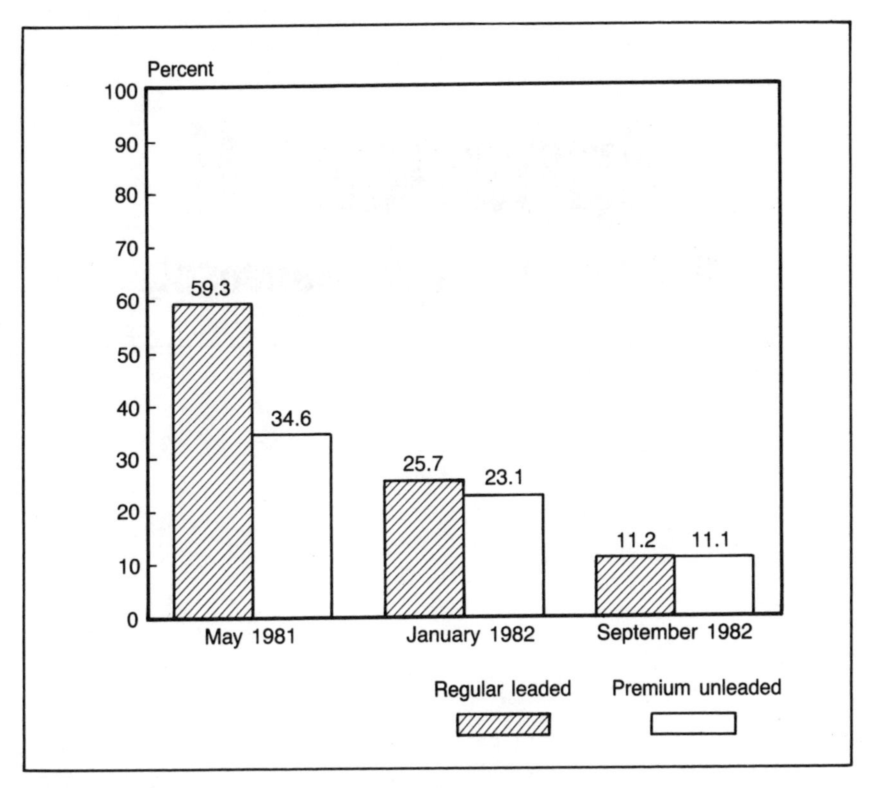

Fig. J.4 *Pemex's declining share of retail gasoline prices (source:* Energéticos, *July 1982 and* Energy Détente, *February 2, 1982)*

Appendix K
U.S.-Mexico interrelationship

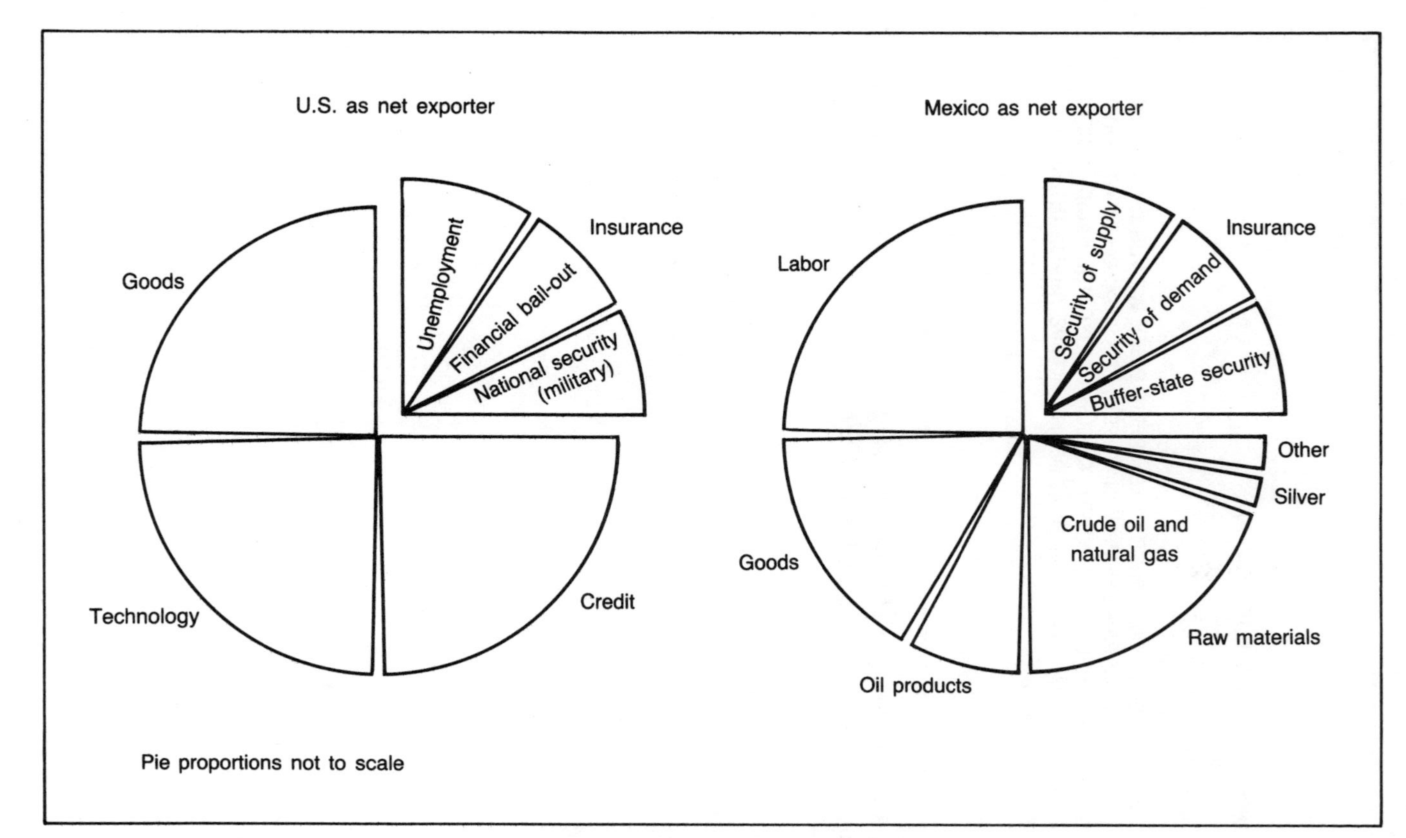

Fig. K.1 *U.S.–Mexico buffer services trade (source: Baker & Associates)*

MEXICAN PETROLEUM COMPANY
120 BROADWAY,
NEW YORK CITY

YBEKV BRAIM GAQQU IPBYR FETSY SFEIE BIBFY WEWUR ENYIH PLOFZ
JIYUV ESOYM OEPAN EYWFE WINAY IKEIH SEAJI QYCOH NEYDJ SHYAE
BUBIB IHXFI IPBEL ORUUD XUVTU QYCOH EWOYD ADEII ACIDL IHXFI
IPBEL FYUNZ ROYKV TNANI TIQIH REEHD HOMVE ARROGANT AHOIC
SVUVE XYRUD

Fig. K.2 *Mysteries of U.S.–Mexico energy relations (telegram to a U.S. oil company from its Mexican office, c. 1920s), (source: Pemex)*

Additional readings

In English: An excellent overview of the management history of Pemex is in George Philip's *Oil and Politics in Latin America* (Cambridge, 1982). Edward J. Williams gives political insights in *The Rebirth of the Mexican Petroleum Industry* (Lexington, 1979). The best, book-length discussion of Mexico's resource base is the Rand study prepared by David Ronfeldt, Richard Nehring, and Arturo Gandara, *Mexico's Petroleum and U.S. Policy* (Santa Monica, California, 1980). An earlier work on *Mexico's Natural Gas* (Austin, 1968), by Fredda Jean Bullard, is a broadly conceived study that deals with both geological and industrial issues. Frank Niering's reflections on the size of Mexico's commercial oil deposits, "Mexico: The Problems Mount," *Petroleum Economist* (July 1982) deserve close reading.

A.J. Bermúdez's memoirs were revised for an English edition, *The Mexican National Petroleum Industry* (Stanford, 1963). The keenest Mexican student of the early years is Lorenzo Meyer, whose *Mexico and the United States in the Petroleum Conflict, 1971–1942* (Mexico City, 2nd ed., 1972; Austin 1977) is standard. Carlos Fuentes, from whom a great deal about Mexico can be learned, wrote an oil novel, *The Hydra Head* (New York, 1978), which insightfully explores cultural and political issues. Industrial and commercial sides of the post-expropriation period are treated in J.R. Powell's indispensable *Mexican Petroleum Industry, 1938–1950* (Los Angeles, 1956). More recent matters are discussed in George W. Grayson's *Politics of Mexican Oil* (Pittsburgh, 1980), Richard Mancke's *Mexican Oil and Natural Gas* (New York, 1979), and essays published in *U.S.-Mexican Energy Relationships* (Lexington, Mass., 1981), edited by Jerry R. Ladman, Deborah J. Baldwin, and Elihu Bergman. A sensitive critique of the López Portillo period, written by a Mexican historian, is Enrique Krauze's "Mexico: The Rudder and the Storm," *The Mexican Forum* (Austin, January 1983). An uncommon discussion of the social and ideological origin and nature of the Mexican State is found in an earlier, neglected work, *Catholic Church in Mexico* (Mexico City, 1965) by Paul V. Murray, who is best known for having been one of the founders of Mexico City College and the University of the Americas. Cover stories on current issues appeared in *Time,* "Mexico's Crisis" (December 20, 1982) and in *Business Week,* "Why Pemex Can't Pay Mexico's Bills" (February 28, 1983). Much of Díaz Serrano's tenure at Pemex may have to be rewritten depending on how much is allowed to come out in the several court cases pending in the U.S. and Mexico; the Crawford case was reviewed in "Mexican Showdown," *Wall Street Journal,* February 23,

1983, p. 1. For keeping up on Mexican political and energy matters, Barney Thompson's weekly *Mexican News Synopsis* (San Ysidro) is helpful, as is Trilby Lundberg's *Energy Détente* (North Hollywood).

In Spanish: The availability of materials in Spanish on Mexico's oil sector is sharply restricted, in part by the Mexican custom that advises that the incumbent president and his major policies are not to be criticized in public by subordinate officers in business, government, or academia. This restriction, however, is not enforced on policies of previous administrations. A constructive review of Pemex management policies and practices under López Portillo was given by Pemex president Mario Ramón Beteta in his Petroleum Day address of March 18, 1983. Other criticism, much of it not constructive, is found in the work of Mexican political cartoonists and satirists, such as that of Herberto Castillo/Ruiz, whose *Huele a gas: los misterios del gasoducto* (Mexico City, 1977) was a mordant critique of the management of the natural gas pipeline project. Such criticism usually focuses on political and ideological issues; for example, the thesis that the U.S. might invade Mexican oil fields is speculated upon in Luis Suárez's *Petroleo: ¿Mexico invadido?* (Mexico City, 1982). Various attempts to demythologize Pemex have appeared, one of them by Luís Pazos, *Mitos y realidades del petróleo mexicano* (Mexico City, 1979). Raul Prieto's highly critical, but undocumented, *Pemex muere* (Mexico City, 1981) was received in great silence in official Mexico. Mexican academicians who publish on energy matters, such as Olga Pellicer de Brody and John Saxe-Fernández, generally restrict their focus to national-policy matters. Representative essays are Pellicer de Brody's "La política de los Estados Unidos hacia el petróleo mexicano, 1976–1980," available through the Program in U.S.-Mexican Studies at the University of California, San Diego, and Saxe-Fernández's "La dependencia estrategica y el petróleo en las relaciones de Mexico y los Estados Unidos," in a book of essays, many of them reprints, published by Mexico's National Science Council (CONACYT), *El petróleo en México y en el mundo* (Mexico City, 1979). Miguel Wionczek, head of an energy think-tank at the Colegio de Mexico, did pioneer work on the history of the electricity industry in Mexico. His "La industria eléctrica en Mexico, 1900–1960" appears in his book *El nacionalismo mexicano y la inversión extranjera* (4th ed., Mexico City, 1977). A collection of Colegio de Mexico essays is *Las perspectivas del petróleo mexicano* (Mexico City, 1979). The well-known essay by Jorge Díaz Serrano "¿En que consiste una reserva petrolera?" is in the CONACYT volume. An interpretive essay, "Significado económico del petróleo mexicano en las perspectivas de las relaciones México-Estados Unidos," by an energy advisor in the President's office, Jaime Corredor, is in *Las relaciones México-Estados Unidos* (Mexico City, 1981), edited by Carlos Tello and Clark Reynolds.

Regarding the Díaz Serrano Affaire, which in August 1983 put Pemex's former CEO behind bars for his alleged involvement in a scheme to defraud Pemex of $34 million, Mexican public opinion is divided: one side accepts the government's case as stated, the other side writes it off as the product of mere in-fighting between political factions. Díaz Serrano's self-defense before the national Chamber of Deputies is vague and philosophical with regard to his tenure at Pemex, but sharply focused when he turns to a technical argument asserting that his constitutional rights had been violated repeatedly (*Excelsior*, July 31, 1983, p. 10A).

Index

D

E

I

J

K

L

M

N

O

V

W

X

Y

Z